Other Fiction by the Author

South American Historical Adventures (as Hernan Moreno Ruiz)

Letters from San Rafael
Return to San Rafael
Confessions of Father Xavier

Science Fiction
The Ice Ship
200 Years Before the Mast
Orion and Other Stories of the Future

Humour
The Innocence of Tom Shipley: Teacher.
Tom Shipley's War: Memoirs of a Weekend Warrior.

Australiana
The Cry of the Currawong

Supernatural Crime (Omnibus Editions)
The Casebook of the Ghost and Detective McNab
More from the Casebook of the Ghost and Detective McNab

The author also has written many works of non-fiction in the Earth Sciences, Environmental Sciences, Teaching and Survival genres as both Print and eBooks.
https://www.amazon.com/s?k=%22Dr+Peter+T+Scott%22&ref=nav_bb_sb

Felix Publishing 2024
email: info.felixpublishing@gmail.com
Print copies available from publisher.

Poseidon's People

Print Edition
ISBN: 978-1-925662-52-8
Digital Edition
ISBN: 978-1-925662-53-5
Author: Dr Peter T. Scott

Registration:
Thorpe-Bowker +61 3 8517 8342
email: bowkerlink@thorpe.com.au

This is a work of fiction. The characters in this book did not exist and the politics of the time have been generalized. Some of the places described are real and are well-known to the author. No disrespect is meant to any person living or dead.

Poseidon's People

Peter T. Scott

First released 2024

Table of Contents

Chapter 1: The Wave

Nathaniel Pierce leant back in his rocking chair on the veranda of his house on the shore front of St. Augustine Beach while gazing over the vast expanse of the North Atlantic Ocean. The early morning sunlight sparkled as a broad band of colours of yellow and red shimmered across its flat, blue plain. There was a gentle breeze blowing in from the east and the early morning sun felt good, bringing the tangy smell of saltwater and a smile to the old man's face. He relit his old brier pipe and took a pull of the rich Irish tobacco mixture which he had imported from a company in County Sligo. It was especially infused with a good brand of whiskey from that country, and it brought back pleasant memories of his early days as a young sea captain on the North Atlantic run. Machinery and other manufactured goods in exchange for timber; that was the trade and a good pipe, and a heavy woollen pea jacket were the only comfort on the lonely Second Watch from midnight until four in those icy cold waters. The warmth of the early morning Florida sun gave him some comfort when he thought of his days as a young cadet freezing on the bridge in the North Atlantic.

He was a thin, wiry man whose leathery brown face had been moulded by sun, wind and age. He had been well over the average height in those days, and even now he would be considered a tall man when he stood up; which was uncommon nowadays with arthritis in both knees. Captain Pierce, as he still liked to be called by the equally old members of this small Florida community, was in his late seventies but, apart from his knees, was in good health otherwise. He had retired from the sea many years ago and had moved south from his beloved New Bedford, that well-known former whaling town in Massachusetts, to these warmer climes. Pierce sometimes grumbled to his poor wife, Marjorie that this was like being marooned 'on the beach', but these were only during the better days of warmth and sunlight; too often his old body reminded him that he had spent far too long on cold and icy seas. Marjorie, who seemed to take everything in her stride had prudently calmed these few outbursts by reminding him of those early homecoming days of a warm greeting and a comforting fire in the living room of their old two-story home near the waterfront of New Bedford. But now in the warmth of Florida's summer and the ease of retirement, he could look upon the sea now as an old benign friend and only the good memories came to mind; the bad memories of turbulent seas with the spray turning to ice before it hit one's unprotected face had mostly passed.

He had watched over the sea most of his life, from those early, exciting days when, as a teenager and raw cadet he had first stepped aboard his first ship. That was a grand day. He had been accepted by one of the bigger, well-known shipping lines as a cadet with the promise that he may one day get his Master's Ticket. Becoming a full captain of his own ship was well off in the forgotten gloom of the future. Eventually through hard work and study he had achieved all of his tickets and certificates and had been given command of his own ship. It had not been much; just a small coastal freighter taking wood from Victoria in Canada down to San Diego, but it was his ship and those early days of being in command and being responsible for everything from freight manifests to minor surgery brought another smile to his lips.

Whether on watch or just catching a few moments of solitude on the bridge wing, he had seen the sea in her many moods; sometimes she would be calm and reassuring, allowing ships and their human crew to travel peacefully across her vast domain and at other times she would vent her anger at these intrusions and would rise up in full fury of wind and water to destroy any who could not tame her. Yes, Captain Nathaniel Pearce knew all of the world's oceans; the Atlantic, the Indian and the Pacific and a few not often travelled by cargo vessels.

Pacific, Ha! He thought, as he gazed upon the broad stretch of this other now peaceful sea; *Peaceful sometimes but angry the next!* he thought to himself with a wry smile on his old, weather-beaten brown face, creased with many a day on deck with the force of sun and sea-spray on his face. He had travelled all over that great ocean, from the cold of its Alaskan waters, through the torrid temperatures of the tropics and into the even colder realms where it became the great Southern Ocean which surrounds Antarctica. Here he had twice sailed the feared Drake Passage between that continent's long peninsula and the rocky, windswept shore of Cape Horn and both times had found its reputation truly justified. There, the wind of the 'Shrieking Sixties' roared like some untamed beast through the narrow separation between South America to the north and the Antarctic Peninsula to the south. The waves, wiped up by this great wind from the west in this narrow funnel, often towered above their small ship which bounced like some small toy to the very crest of each wave to show the luckless crew the vast, grey boiling vista of that wild sea. Then, just as suddenly, their small ship would plunge almost vertically into the depth of the watery trough surrounded by menacing grey-green walls capped with thin white crests and the hope that they would once again be resurrected to the violent heights of wind and water above. Inside the ship, it's the poor human cargo would be thrown about like mere marionettes unless they tied themselves to their bunks in cabins, wet and dank, or lashed to the wheel on the bridge; fearful eyes on the wildly swinging Lubber Line in the compass binnacle showing their direction and course between unforgiving shores. It would be a brave man who would go out onto the ice-covered deck at those times, even lashed to the tenuous lifeline which ran across the ship's slippery deck plates.

Now Captain Nathaniel Pearce could relax, sit and rock gently back and forth in his beloved old Salem rocking chair and enjoy the view of what was now a very peaceful Atlantic Ocean. Hardly a ripple on its sparkling blue surface. Just the reflections of the early morning sun on the water. He was at peace; especially now that he had his daughter June and his little grandson Tony with them for a short stay. When he looked back down from the sea to the long, stretch of sand across the narrow road and the low dunes in front of his house to where June and Tony played, he was glad that his daughter was happily married to another man who also loved his sea. Unfortunately, not a sea captain like himself, but a world-renown oceanographer who held some senior rank in the country's Naval Reserve. It was disappointing though, that his daughter and grandchildren still lived well to the north near their old home in New Bedford. But for a while, he was content that his daughter had come to stay for a short time whilst her husband, Dr. Terence Breck had travelled to Europe where he had been invited, as a Visiting Fellow, to address some international symposium on climate change at the University of Edinburgh.

June had long ago resigned herself to her husband's many trips away from home; it was not always possible for him to do his research at his office at Woods Hole Oceanographic Institute just across the waters of Buzzard's Bay. If he wasn't going south to the Naval Ice Center at Suitland, near Washington, he would be out at sea doing some research in some cold part of the world's oceans. Recently, he had only just returned from an extensive stay in east Antarctica with one of the Centre's research vessels looking at waters along the Antarctic Peninsula. She naturally was somewhat glad when he was asked to travel to Edinburgh University to give a talk about his findings in the Antarctic; it would be a lot safer there and at least the three of them had been able to meet briefly at JFK airport. He had gone on to Scotland where she and their son Tony had taken a domestic flight to Jacksonville and then on the short bus ride to St. Augustine.

Nathaniel Pierce rocked back on his chair and once more looked up at the great expanse of sparkling blue water in front of him. It was lucky that he had been able to have this house built so that there was a fine view of the sea. Well, he thought, it had been mostly Marjorie's work; she was the brains of the family. Always was! He went to sea, and she stayed at home and managed the accounts of their small shipping company of two small freighters as well as the family of three girls. She had planned this retirement well before he even knew that he was going to retire. That was her way. He laughed quietly when he thought of the way she handled such important decisions. They had flown south and spent over a week looking at houses and land; my God, what pests some real estate people could be! None of the homes they were shown suited them at all; they were either too far back from the sea and could only provide a view of thick mangroves or were those expensive horrors favoured by the modern breed of architects and builders who thought only in terms of flat-rooved rectangular boxes. They reminded him of those new-fangled shipping containers that crowded out the decks of those over-sized ships commanded by computers rather than men. These monstrosities however, he simply thought of as large fragile boxes which somehow had been cast up onto the manicured beach estates along the Florida coast.

But eventually, they had found a good block of land well out on its own small peninsular, separated from the low-lying sand dunes and the small beach beyond by a narrow bitumen roadway which led into town. They had built a small, two-story house which could be managed easily; not unlike those old, pretty, coastal cottages back at New Bedford. Despite the real estate agent making some small comment that one day there might be a problem with sea-level rise, they had purchased the block anyway.

Sea-level rise! Rubbish! He thought. He had watched the tide come and go for most of his life and did not believe in that global warming nonsense put about by so-called 'Investigative Journalists' and social media 'experts'.

Marjorie had been active again in the decision-making and had managed to find a local builder who could construct an older style wooden home of two stories with a good, wide veranda in front and space for a garden behind. The house was built with enough time for them to have their favourite pieces of furniture and personal belongings transported on the long road trip from New Bedford to St. Augustine.

Yes! He was happy here as he looked down and watched his daughter and his little grandson playing on the small beach in front of his house. He looked past the dunes and the beach to the long line of the horizon separating a cloudless blue sky from the sparkling calm of the early morning sea.

His peaceful gaze was suddenly interrupted. There was something amiss in this peaceful view. Something was not right! He had spent too many years on the bridge of ship and had an eye for detail and anything which was unusual always caught his attention. He took off his rimless spectacles and wiped them on his shirt sleeve. His attention had been caught by a long, low line parallel to the horizon but well out to sea. He slowly eased himself out of his old rocking chair and ambled across to where he had set up an old brass telescope mounted on an even older tripod. He would sometimes delight in looking at the magnified details of some passing ship as old habits die hard with Captain Nathaniel Pearce.

He turned the old telescope around, slowly bent down and looked through its tarnished brass eyepiece. Adjusting the small knob on the barrel of the telescope's long tube, he brought the mysterious line into clear focus. It was a long wall of water stretching across the horizon and coming rapidly towards him. He stood up and wiped his old eyes and then looked again.

No! Impossible! he thought. *Not in these waters!* He had seen such a sight only once before on another calm day off the Sunda Strait between the Indonesian islands of Java and Sumatra where it connects the Java Sea to the Indian Ocean. Then, in the deep water, his small ship had been able to rise up with the huge wave and then slide safely down to the flat green water behind it. There had been no storm then nor now and certainly no warning of any earthquakes ashore. Such disturbances were not known in this part of the world, especially not off the Florida coast.

But it was exactly what he had remembered all those years ago! It was a tsunami front coming directly towards the Florida coast. He knew from tales he had heard from some of the older mariners, whose ships had once been anchored safely in their harbour but now rested on the shore well inland or on the bottom of the ocean, that such a wave suddenly built up into a fearsome green wall of water. It crushed everything in its path as it roared relentlessly up through the shallows and onto the land.

Nathaniel Pearce pushed the telescope quickly aside and rushed as fast as his old legs could go to the front of his veranda and yelled with all of his strength to his daughter and grandson playing on

the beach just across the dune: "June! Run! Run for
of your life!"

Chapter 2: A World in Crisis

Dr. Terence Breck felt happy in himself but disappointed in the collective inactivity of the symposium which had just concluded. He had been invited as a minor guest speaker to the University of Edinburgh to give a Southern Hemisphere perspective on what was primarily a symposium about Northern Hemisphere currents and climate change. As he walked down the broad, flat steps which led out from the Grant Institute and into James Hutton Road he wondered what the future held for scientists like himself when complacent governments and others with vested interests chose to ignore the overwhelming scientific evidence for climate change.

The building which he had just left was once called the 'New Geology Building' by its earlier students when it was built as the Grant Institute of Geology in 1931. So, it was appropriate, that the road on which it stood, had been named after the founder of modern Geology, James Hutton. Its classical auditorium had also been the venue for what Breck had thought would have had the potential for being the definitive international teleconference on current research into climate change and global warming. He had been disappointed that the only conclusion which had finally been reached was that there should be another meeting in the following year. There was a cold wind blowing down the narrow tarred and grass-bordered road which ran across the front of the old buildings. It was coming on to summer now, but the winter had been bad, and now it seemed reluctant to give up. Breck pulled up the collar of his coat and felt that perhaps all of the proceedings of the conference were simply yet another waste of time.

As a Visiting Scholar, he had been asked at the last minute to speak at this, the International Symposium for Oceanic Research into Climate Change, the most prestigious meeting of the world's influential scientists to date and considered by some to be the last hope for solutions to global warming.

Breck had been happy to travel from his home and research base in the United States to Scotland to present his own paper at the symposium. It had been several months now since he had published his latest findings on the research he had undertaken over the last five years in the Southern Ocean and the Weddell Sea in Antarctica. He had never visited Scotland before but had always felt a strong bond to the country where his ancestors had lived before emigrating to America in the 19th Century.

Breck worked for a branch of the Antarctic Division of the U.S. Naval Oceanographic Office (NAVOCEANO). His research had been in the changes in bottom sea floor currents and their changes in salinity, as opposed to that usually measured near the surface. His main premise was that this was a better measure of global warming, and his findings had suggested that the rate of ocean temperature increase was alarmingly greater than had been predicted by more conventional modelling. He also had many reservations about the computer modelling, which was currently being used, especially when the algorithms were designed by computer geeks who had hardly set foot on any oceanographic vessel.

There had been a strong and often lively discussion during this tele-conference, mainly from some of the local and televised academics who were guest speakers, but also from several of the University's students and members of the general public who had crowded into the auditorium. There had been some loud and often rambling questions from many of the general public and the usual one-sided and aggressive arguments from a small group of so-called 'climate activists' who had taken over some of the seats in the auditorium's upper gallery.

The Chairman of the symposium, Professor Angus Campbell had opened the symposium with his customary good-natured welcome and the joke about climate change eventually leading to Scotland being the new Riviera. As a prelude to the symposium's topics, he had gone on to summarise the current understanding of the natural causes of climate change. These included: changes in the output energies of the Sun; variations in the orbital properties of the Earth; how the Sun's rays were absorbed by the atmosphere; or reflected of the surface as the albedo effect, especially at the ice caps; the effects of warm and cold ocean currents; plate tectonics and continental drift; effects of volcanic eruptions; and the quantity of greenhouse gases in the atmosphere such as water, methane and carbon dioxide. He concluded with the current evidence from the Mauna Loa Observatory in Hawaii that the increase in carbon dioxide, as a result of humankind's use of fossil fuels, had risen to over 430 parts per million; a giant increase in this Green House gas emission from that of about 280 ppm prior to the eighteenth century. This introduction had been given in the Professor's usual genial manner; a style which usually went down well with his students who because of their affection for him and his rotund shape often privately referred to him as" Big Mac".

In contrast, the first guest speaker had been one Dr. Isiah Black from the Royal Greenwich Observatory who was a very dour speaker and who delivered his talk with total disinterest and generally in a sleep-inducing monotone. Moreover, unlike the good professor, his detailed description of how climate change could be affected by changes in the Earth's orbit was full of technical terms which even many of the more technically minded of the audience found difficult to digest. He described in some pedantic detail the theories of the late Serbian geophysicist and astronomer Milutin Milanković who, in the 1920s, hypothesized that it was natural variations in the position of the Earth around the sun which contributed the main effects of its climate history. These variations included the eccentricity of the Earth's orbit, the planet's axial tilt, and its precession or axial wobble. These all combined to produce cyclical variations in the intra-annual and latitudinal distribution of solar radiation at the Earth's surface, and thus strongly influenced the Earth's climatic patterns. At the end of this talk there were no questions from the audience and great relief when Professor Campbell announced that a generous morning tea was to be had in the foyer of the school.

There was nothing really new from the speakers following this short recess; a lesser-known member from the United Nation's Intergovernmental Panel on Climate Change (IPCC) gave the usual warnings to governments about the need to reduce greenhouse emissions immediately and a spokesman from the British government gave a long-winded and often rambling set of assurances that the current government, unlike those of the previous administration, were working towards some vague final conclusion.

The morning session, having finished promptly at noon, had been followed by a sumptuous lunch at the University's main Refectory, putting the delegates in a more relaxed mood. The afternoon session of the symposium continued with talks, both from speakers within the auditorium and a few via the tele-conferencing network, on such topics as the future of the world's energy requirements, current research on alternative energy and the need for political cooperation. There had been many questions from the floor and from the teleconference delegates but very few definitive answers.

Breck's talk on his research in sampling the ocean floor salinity along the warm Brazilian Current and into the colder waters of the Antarctic Peninsula, had been the last for the day. His conclusions had startled some of the more conservative delegates who were comfortable in their own belief that global warming only occurred in the Northern Hemisphere and that its rate was slowing down.

At the end of his talk, Breck had taken questions from the floor; mainly from ill-informed people who categorically claimed that parts of Antarctic had not been affected by climate change. Breck had quietly explained that Antarctica did suffer these effects, especially glacial melting along the Antarctic Peninsular where he had worked, but he did concede that eastern Antarctica suffered far less than the majority of the continent. He had called up one of his graphs to show that Antarctica was indeed subject to climate change, reiterating that latest research showed that melting was also occurring more from the base of the ice sheet and less from its surface, so did not easily show up on satellite images. This had had the desired effect on the sceptics and as a result, the response to his talk had been generally well received.

There had been one question only from one of the teleconference delegates, a Dr. Yuri Ziminov from the Siberian Academy of Natural Science in Krasnoyarsk. His question really was a way to outline his own theories and a recently published paper; himself not being an official speaker. Ziminov's manner seemed to be one of extreme agitation and urgency and his clipped manner of speaking and intense furtive gaze reminded Breck of some old Cold War caricature of a Soviet anarchist. He had asked Breck about the rate of snowfall currently occurring in and around the Weddell Sea in Antarctica which was becoming more free of ice. Breck apologised for the fact that this was a little out of his research area, that being counter current salinity measurements. He did, however, outline briefly some of the findings from the annually logged ice core data from the Ross Sea in western Antarctica, named the Roosevelt Island Climate Evolution (RICE) ice core study. This data had suggested a trend over the past 2700 years that the eastern Ross Sea was indeed warming, and there was considerable melting of the sea ice in recent times due to modern global warming. However, it had been noted that there had also been some increased snow accumulation following the decrease of some surface sea ice.

Dr Ziminov seemed to get quite excited about Breck's description of increased snow accumulation in the land near the eastern Ross Sea. The excitable Russian seemed to be very pleased with himself, almost triumphant, as though he had finally found some support for his work. He had then ignored the rest of Breck's answer and had continued to enthusiastically outline his own theory. He maintained that the opening up of the sea ice in the Arctic Ocean, and in particular in his own research area off the coast of the Kara Sea in the far north eastern part of Siberia, was leading to increased snowfall on the adjacent mainland of Russia. His research had shown a steady increase in snowfall in north eastern Siberia as the amount of open water in the Arctic Ocean increased. In suggesting this trend, he had invoked the old theory of Maurice Ewing and William Donn of 1956, which suggested that in the past geological time, removal of sea ice from the Arctic Ocean had led to the formation of massive ice sheets which had then spread across northern Europe and the Americas. Ziminov had concluded that current global warming and the subsequent removal of ice from the surface of the Arctic Ocean would lead to a new Ice Age in the very near future.

Chapter 3: The Sea Wolf

Dimitri Alexandrovich Volkov opened the door and walked onto the spacious bridge of the Russian nuclear-powered icebreaker the *Vasili Brusilov*. The atmosphere was hot and stuffy and full of the acrid odour of the cheap Makhorka pipe tobacco which the captain like to smoke in his old brier pipe.

"Permission to enter the bridge?" he said as an after-thought as he walked past the captain who was examining the ship's ice radar. Captain Ivan Grigorievich Khovrin briefly looked up and gave a cursorily grunt to the newcomer who had just come onto his hallowed bridge. The captain disliked Volkov but tolerated his entrance and indeed his very existence on board the icebreaker. After all, Dimitri Alexandrovich Volkov was the owner of the ship which was just one of many which he owned as chief stockholder of Sea Wolf Shipping Lines OAO[1].

Captain Khovrin was a sea-captain of the 'old school' who had come up from the ranks of Cadet Seaman to command his own ships. In his early sixties, Captain Khovrin looked every inch the Russian sea captain. He was of average height but stocky, with a broad, weather-beaten Slavic face which, like the seas he had travelled for those many years, could be calm and beset with a wide smile, laughter and sparkling blue eyes or stormy with a face that clearly showed his anger and dissatisfaction. Usually, he was not a man to hide his emotions except in the presence of Volkov but his crew loved him for his direct approach and his fair leadership; he was a true sailor, and his ship was generally a happy one. But he disliked Volkov because he was his complete antithesis; untrustworthy and secretive like many of the corrupt oligarchs he had had the misfortune to

---

[1] OAO in Russian refers to Открытое Акционерное Общество or Public Joint-Stock Company, a public company of shareholders.

meet in conducting business for his ship. To Captain Khovrin, his employer always reminded him of some predatory lean and hungry animal; both lean in physical appearance and hungry for business opportunities and power. The captain had never liked any representative from company headquarters. 'Suits,' he usually called them with usually a mock spitting towards the floor; unthinking and pretentious individuals who thought that shipping cargo was simply a matter of full manifests and economics. They had no idea of what it was like to carry their precious cargos across unforgiving seas.

The new arrival and owner of the company, Dimitri Alexandrovich Volkov, thought little of stocky captain on the well-ordered and tidy bridge of this, the company's most modern ice-breaker. He thought of the captain as simply an uncultured crude individual; very much the sea-going peasant but useful for the present until he could replace him with one of his more loyal younger captains. It was only because his late father had put much into Captain Khovrin's reputation within the company that he had retained the old man on his ship.

Walking past the now silent helmsman who knew better than to show anything other than an intense concentration on the helm, Volkov strode over to the end of the bridge wing where the double-glazed and heated windows gave an excellent view of the sea. He looked out at the grey expanse of the Arctic Ocean; he never tired of looking at such a seascape as he felt that he was born to the sea and indeed to command these very waters. His grandfather, Mikhail Volkov had gone down with his torpedoed ship in 1943 as it approached Kola Bay heading for Murmansk in the Great Patriotic War. Then his father had built on the considerable legacy of the old man to build up the Sea Wolf Shipping Line, an aggressive cargo-carrying line of merchant ships and ice breakers which operated out of Murmansk and generally plied the North Atlantic routes for any cargo that was available. Now it was his eldest son, Dimitri who ran the business after a very brief career as a Cadet who took advantage of the recent opening of the Arctic

Ocean during the summer to forge new shipping routes between Europe and Asia and break the crippling sanctions which were now imposed by the Western nations on Mother Russia.

Volkov heard the door to the bridge open and turned to face his old friend and Head of his Special Operations Bureau, Sergey Sokolovich Ustrashkin. Ustrashkin was at least a head taller than his employer and powerfully built; his shaven head and a scar across his temple only added more menace to his normally grim face. They had met doing their compulsory period of conscription many years ago during service in Chechnya and had become firm friends. Volkov had resented every day of this service and cursed his father for not 'buying out' his commitment like many of his privileged friends, but he had come to like the simple and practical nature of the tough, ex-street fighter from the southern Russian sea port of Novorossiysk on the Black Sea. Ustrashkin, for his part, admired the intelligence and education of the privileged northerner from Murmansk and soon learned that following his friend had many advantages. Whilst Volkov's military career only lasted the usual minimum period of his country's 'Universal Military Obligation', Ustrashkin had remained in the army where he had found a socially approved outlet for his naturally aggressive behaviour. Continuing on and seeing active service in several of Russia's border wars, he had eventually joined the Spetsnaz GRU, the politically orientated Special Forces unit, becoming a sergeant. When he finally had had enough of the demanding military life and the dangers of combat in the south and east of the country and as a mercenary in Africa and the Middle East, he had left the military and had sought out his old friend for a job.

Volkov was naturally delighted to employ his old friend and because of his military training and experience, had made him Head of his new Special Operations Bureau. For Ustrashkin, this position was ideal; high-paying and simply a civilian application of his old duties in the Spetsnaz GRU.

He had a healthy budget and employed a hand-picked team of former soldiers and GRU members who could carry out the necessary corporate intelligence surveillance of Volkov's rivals. Several of this team were now on the icebreaker as Ustrashkin, being a suspicious man, always travelled with his own well-armed bodyguard.

Occasionally, the work of such a security team also required 'special operations' against some of Volkov's competitors. Thus, Ustrashkin's private intelligence network operated far beyond corporate Russia and international law and into the wider world of commerce, the media and even academia where new research could be an advantage or disadvantage for his employer's wider ambitions.

"It's a fine view, is it not Sergey?" Volkov said, with a thin smile on his usually gaunt face, sweeping his arm out enthusiastically and turning once more to look out from the bridge window. Ustrashkin looked past his superior's shoulder at the dark blue-grey Arctic Sea merging in the distance with a cloudy, light grey gloom where the sky should be. In the immediate distance, the dark grey of the sea surrounding their slowly moving ship, was filled with many white and jagged pieces of floating ice. These were mostly all similar in size, being what old hands in the Arctic Ocean called 'bergy bits'; large chunks of snow-capped white ice, sometimes with a blue opalescence at their base and ranging in size from about a metre to five metres above the surface of the sea and covering on average about two hundred square metres - about the size of a small house. Closer to the ship, he noticed smaller 'growlers', which were mostly flat-topped ice slabs, often awash and difficult to see; a real danger to shipping if they were of any larger size.

Although he was confident in the strength and power of the ice breaker, Ustrashkin had never felt happy about a sea full of ice, most of which was below the surface and capable of ripping the thin hulls of ships apart, even careless icebreakers.

Volkov on the other saw only the openness of the sea between the ice and thought about the onset of summer which would make the Arctic Ocean more amenable for the passage of his cargo-carrying ships.

"I'd much prefer the warm sunlight on my native Black Sea and its blue sky above, Dimitri Alexandrovich." Ustrashkin replied, zipping up his jacket tightly more out of habit than what was necessary as the temperature on the bridge was excessively warm.

"Ha! Don't worry Sergey. With more global warming you will soon prefer my northern climate to your hot, sweltering Black Sea weather. But no matter! It won't get that bad. Meanwhile we can enjoy the benefits of my ships taking their cargos across the Arctic Ocean to many of my Asian customers despite Western sanctions and the new Baltic alliances in NATO. Now that my company is publicly listed, and with the many shares we both have in Sea Wolf Shipping, we can retire in a few years to our dachas on the Mediterranean and let those ninnies in the United Nations cry over the melting of the northern ice cap".

"I am still worried you know, Dimitri Alexandrovich. There are some people who do more than just talk about these changes in climate which are helping the company making these profits." Ustrashkin said in a low voice, not wishing to anger his employer. Volkov turned and looked at his friend whose tough face showed some real concern.

"Ha! We don't need to worry about the rantings of those so-called environmental groups; no one else takes them seriously anyway. And as for those academicians in their ivory towers, well! Let them publish or perish! They have little influence on the politicians who prefer to maintain their positions of power, take our company's royalties and keep their 'home fires burning' as they say!"

"Yes, sir" Ustrashkin replied with the appropriate amount of diffidence "But my contacts in the world of those academics and who watch such things carefully, has alerted me to a recent conference in Scotland which they monitored as part of their usual surveillance schedule. There were reports given about the melting ice caps with references to recently published articles which may cause us some problems in the future."

"Don't worry, my friend. Global warming, the melting of the northern ice cap, and our future profits are going to go on for a long time. We will be long gone, and our children's children will still be reaping our company's profits before anything else will change." Volkov said with a smile on his broad face. "But! If your contact's fearful academics are possible threats to our considerable share portfolios, then I am sure that your team will know how to remove them in the usual manner.

"Yes, that is true Dimitri Alexandrovich. I will see that it is done as soon as possible; an unfortunate accident perhaps?"

"A good precaution Sergey. If the company's shares fall, then so does your comfortable retirement. See to it." Volkov said with a scowl as he turned from the window and walked quickly across the bridge.

Chapter 4: An Unexpected Invitation

Breck walked quickly along the roads south of the university which skirted The Meadows hoping to make it to his lodgings before the sun set. The blue bitumen - capped roads were flanked on either side with austere stone and brick buildings, built together as though to deny any sunlight to their forlorn streets. They were usually of four stories with a shop of some kind occupying the ground floor. Eventually he came into the open freshness of Mayfield Road with its many tall trees and flowering gardens and headed in the direction of the guest house where he was staying. Evening was just setting in, and the sunlight, now faintly coming from over the Pentland hills to the west, made long shadows from the many trees along the road now turning into their many colours of early fall. He never liked hotels, especially the large multinational ones. They were always the same as those he avoided back home, the same standard rooms, the same mass produced foods and the same insincere smiles at the reception desk. He much preferred to stay in local accommodation and share the life and culture of the places that he visited. In a small guest house, he was usually treated like one of the family and enjoyed the feeling of comfort in a truly domestic setting. Here in Scotland, the country of his ancestors, who had fled poverty in the Highlands during the great Clearances over two hundred years ago, he felt a little like he was really home. He also revelled in the traditional Scottish cuisine served at the guest house; good savoury porridge on a cold morning and a hearty steak pie for dinner often served with neaps and tatties.

He opened the small white picket gate within the hedge which, like most of the houses in Mayfield Road, seem to be continuous along its length as a kind of protective barrier between the old individual Georgian stone buildings common to the district and the busy, noisy road.

Breck walked up the steps to the old wooden front door, opened it and went into the darkened, wood-panelled hallway. The first door on his right was the front parlour, now generally used as the main meeting room for all of the guests in the house. Its bay windows, which always had their curtains open, made the room light and welcoming especially now, with the late afternoon sunlight bathing the room with a comforting red glow. There was the comforting smell of 'old world' curtains and lounge chairs although the room was always kept spotlessly clean. This was enhanced by that of the small open fire burning in the simple iron grate of the ornate fireplace. The hot coals in the fireplace glowed brightly like many small, misshapen jewels as a warning to the chill of the rapidly approaching Edinburgh night.

"Och! ye'r hame noo, Dr. Breck! Tis aff tae be a cauld nicht soon 'n' tea wull be duin at six thirty as usual." Said Mrs Jamison, who owned the guesthouse and had come over from the other front room which she used as an office and private sanctuary. She was a motherly sort of woman who always seemed to have a soft look in her eyes and a smile on her ruddy face. The sort of woman who would see good in everyone but would know exactly how to handle any situation good or bad.

"Och an' thare wis an important lookin letter cam this mornin' efter ye hud left. "She continued, "Ah pat it oan yer bed whin ah wis making it up. Ah will be getting tea soon. Ah hawp that ye hud a guid day".

With that she left the room and walked back out into the long hallway where she switched on the lights and walked down towards the back of the house and its little kitchen where her middle-aged daughter, Jenny was starting to prepare the evening meal.

"Thank you, Mrs. Jamison." Breck said after her and walked over to warm himself by the fire. He took out his cell phone which he had switched off during the symposium and noticed that he had a miscall from his wife, June. He pressed the number and made the call back, knowing that it would be only about midday in Florida. The number answered.

"Hi, darling! How are things?" he asked, happy to be able to make a call to his wife.

"Dreadful, but we are all still alive!" came the firm, typically no-nonsense reply. June then went on to tell her husband about the tsunami which had crashed across the small beach and up into her parent's home above the small set of dunes. Breck was somewhat shocked to hear this news. "A tsunami! In Florida! Are you sure? Are you and Tony O.K? Are your folks alright? he said with some alarm.

"Everyone here is OK, darling. Just a little water damage to the first floor but the sand dunes are gone, and we are still a little shaken. I've been married to an oceanographer for long enough to know what a tsunami looks like you know; the sudden long straight line of the wave which builds up as it comes up into the shallows with the force of the entire ocean behind it. It was a tsunami alright. It came across the road and up into the house. Luckily Dad saw it coming and got Tony and I off the beach in time. Don't worry, it's all over but I would like to stay on here for a few weeks to help Mum and Dad get over the drama of claiming insurance and getting the place back to normal."

"Yes, of course, dear. The symposium is over, but I wouldn't be much use coming back home to an empty house in New Bedford at this time of year."

"Well why don't you kill some time in Scotland for a while? You've always wanted to dig up some ancestor or two. We'll be O.K. here. We have a functioning house, and the power stayed on, so we are quite comfortable upstairs. I doubt that there will be another tsunami here for some time, the local news is full of the story trying to calm everyone down. There was a spokesman from NOAA[2] on CNN news this morning saying that

---

[2] National Oceanic and Atmospheric Administration – the US Government's oceanographic and meteorological agency.

such an event was in the 'very low' category, but they could not account for its cause."

"Well, if you are sure about this? I would much prefer to stay here than fly all the way home. I am sure that your folks don't need a worried oceanographer stomping around underfoot so I'll take your ever wise advice and kill some time listening to bagpipes."

"Ugh! Give me the tsunami any day than having to listen to your beloved bagpipes!" June said with some disgust. It was perhaps the only preference which she and her husband did not share.

Breck changed the topic of conversation from his wife's pet hate of bagpipe music and described the day's proceedings at the symposium. He said that it had been yet another Climate Change waste of time, but he had been happy with the reception of his own research in the Southern Hemisphere and its possible comparison to what was now going on in the north. He said goodbye to June and little

Tony who had come onto the phone and rang off after joking about the so-called 'question' from Dr. Ziminov saying that when the next Ice Age came, they would possibly have to move permanently to Florida.

He put the cell phone back into his jacket pocket and climbed the stairs to his room at the end of the corridor. On the thick floral quilt of his bed was a large, light blue envelope with a small trident in its top left-hand corner. Curious, he opened it up and found a glossy brochure and a letter on the same pale blue – coloured paper and trident motif. It read:

"Dear Dr. Breck,
I would like to personally invite you for a short cruise along the Norwegian coast, possibly into the Arctic Ocean and then on to Svalbard and a flight back to Edinburgh from Longyearbyen. Whilst I was not able to attend your talk at the Edinburgh Symposium in person, I have read your recent paper on your work in Antarctic water and my people had access to the Symposium's teleconference and reported most favourably on the topic of your research.

I hope that you might accept the position as a paid Guest Oceanographer on our ship, the Mystery of the Seas, on this cruise. It is mostly by invitation to a select clientele and a limited number of paying passengers and is concerned with changes to the Earth's climate with specific reference to its effects on the oceans. Our ship is now outfitted as an ecotourism vessel but was once a former research ship, fully equipped for oceanographic and other scientific research. It has been refurbished to carry sixty passengers and crew in comfort. You would be one of three Guest Speakers on board, and it is hoped that you would be able to join our cruise at such short notice. Our ship leaves Leith Cruise Terminal, Edinburgh, at 1000 hours this Wednesday, and our voyage will be of approximately fourteen days' duration. As an American, you do not need a visa for Norway only your passport is required at Svalbard and all travel arrangement and costs will be taken care by us.

Yours sincerely,
Filip Athanasios
Captain,
MV Mystery of the Seas.

Breck read the letter a second time then took up the glossy brochure. It showed a small ship with a blue hull and white superstructure; it looked more like a research vessel than a cruise ship. Indeed, as he read further, he found that the ship had been built in the Hellenic Shipyards, Greece, as an ice-strengthened research vessel but had been more recently converted to what the glossy called a 'boutique ecotourism ship'. Photographs showed rather well-appointed double and single bed cabins, a well-stocked bar and lounge and an extensive dining cabin. The glossy also stated that passengers would have an opportunity to travel along the coast of Norway stopping at several ports, and that there was a planned series of talks and practical demonstrations of scientific research concerning climate change during the voyage. These included talks and activities in meteorology, marine biology, climatology and oceanography.

This cruise seemed very attractive to Breck who now had time to spare and perhaps some personal research into many of the activities brought out in the Symposium would be interesting. The cruise was planned for only two days hence, but he had already packed his few belongings and would only need to rearrange his flight home and tell June of his sudden new adventure; something which he had often done and which she had tolerated for many years.

There was the usual information about how to join the cruise at the bottom of the brochure and having made up his mind, Breck dialled the local telephone number and gave his acceptance as a guest speaker on board the *Mystery of the Seas.*

Chapter 5: The Mystery of the Seas

Breck said farewell to Mrs. Jamison on the front step of her lodgings and walked down to the corner where he was able to hail a cab. It was only a short distance through the north eastern suburbs of Edinburgh to the docks and the cab soon pulled up in front of the small, unpretentious, single story building of the cruise terminal.

There were few formalities of entry here; just the usual friendly passport and ticket check and Breck was able to walk through the small building and out onto the dock. Looking to his left over towards the main Entrance Basin he saw the former Royal Yacht Britannia moored at what the customs agent had said was the Whiskey Wharf. *Gin and tonic!* Breck thought was probably more appropriate.

There was a much smaller ship moored along the cruise terminal wharf to his right. Breck put down his small bag and gave the ship a thorough inspection. He liked what he saw. Breck had always had a fascination with ships even as a boy and his years as an oceanographer and officer in his country's Naval Reserve had brought him into contact with many ships. Some were huge and some were small. Some were ugly and functional, and others were beautiful and decorative. He could easily understand why professional seamen though of ships in feminine terms. The small ship moored at the dock seemed to be both beautiful and functional. It had a narrow hull which came to a sharp point at the bow and the whole hull from waterline to gun'le was painted a dark blue. Above this, the four decks were painted in a brilliant white which contrasted with the single red funnel with its trident logo, red trim around the top of the bridge and the two orange lifeboats on their davits amidships. His professional eye was caught by the equipment located on the stern deck of the ship; two gantry cranes on each side painted dark blue, a neatly stacked rack of Zodiac inflatable boats and a white windlass designed to drag oceanographic apparatus behind the ship. There was also a small, square structure which he assumed would hold the equipment for launching a weather balloon. The whole effect reminded him of a small, compact version of some of the larger naval oceanographic on which he often sailed. In all, the *Mystery of the Seas* seemed to be a beautiful and practical vessel.

There was a small gangway coming down to the dock from a small entry port in the gun'le. Along it was a canvas strip bearing the name of the ship and its trident symbol. Feeling somewhat happy in himself with the freedom of a new adventure about to start, Breck stepped up onto the barred aluminium ramp of the gangway and climbed up to the main deck.

On the deck he found the usual passenger x-ray scanners manned by two tall security guards who both gave him a pleasant smile who asked him to put his bag on the small conveyor belt and then watched him closely as he went through the scanner gate. As he walked into the main lobby of the ship, he was somewhat surprised to meet a very beautiful young woman wearing the white slacks, shirt and epaulettes of an officer. Her impact on his first impressions were devastating. She was very tall; almost as tall as himself which, at six two was generally considered tall for a man. She was slim and her uniform fitted well over her shapely figure. Perhaps the most striking feature about this young woman was her face. It was very beautiful with high cheekbones and full lips which were now in a broad smile revealing a set of almost perfect white teeth. Her skin was well-tanned and contrasted remarkably with her large, light blue eyes. Her long blonde hair was neatly tied in a ponytail which ran down of her back from below her white officers' peaked cap. It was then that he noticed that her epaulettes had the three gold bars on white of a First Officer.

"Good morning, Dr. Breck!" she said, stepping forward with that captivating smile and an extended right hand. "I am First Officer Thalassa. Welcome aboard the *Mystery of the Seas*. I am sorry that we have to provide the usual security welcome, but unfortunately, we live in an untrustworthy world"

Breck took her outstretched hand and for a while was not too sure of whether he should kiss it or shake it. Her maritime marine rank and responsibility was well above that of his Naval Reserve rank. After a brief moment of unsure hesitation, and considering her rank, and he took her hand and shook it lightly. He then looked awkwardly around as if to see how the rest of his dream would unfold. She laughed lightly at his obvious discomfort and beckoned him into the open door of the ship's main foyer. "I will have a steward take you to your cabin to freshen up, but I will be along shortly to take you to Captain Athanasios who would like to meet you."

She gave Breck another engaging smile and signalled with her hand to a steward standing at a short distance away.

"Goodbye for now, Dr. Breck. I will see you shortly." With that she turned and walked smartly through the open doors of the companionway. The steward walked over and gave Breck a welcoming smile and took up the bag which he had placed upon the deck. "Come this way, sir." He said with a foreign accent which Breck could not place. He had noted that the steward also was blonde and blue-eyed and that his skin was the same creamy light brown colour of the First Officer.

Breck followed the steward over to a bank of small, silver metal elevators where the steward pressed the button to go up. The doors opened to reveal a small elevator completely mirrored on all sides except for a small panel of buttons to the right of the doorway. Button three was pressed and the doors closed. When they opened again, Breck found himself in a small foyer wallpapered in light blue with faint images of all manner of fishes. The steward turned and walked off to a companionway on the starboard side of the ship. Like many narrow companionways within a ship, there was subdued lighting and brightly painted cabin doors on either side. A short distance down the companionway towards the stern of the ship, they came to a cabin door labelled '3A'. The steward opened the door with a small key card which he handed to Breck and walked inside. He put Breck's small bag on the floor of the cabin and quietly refused a tip when Breck offered it, turning and closing the door gently behind him. Breck stood in the middle of the cabin with its dark blue patterned carpet and looked around at his temporary new home.

He had often felt the joy of curiosity when entering a new set of quarters on board a ship. There were all the details to take in and how functional they may be. The cabin was small, but adequate for one person with a single, large porthole high on the bulkhead above a single bunk bed. The walls of the cabin were also painted in a light blue but without the decoration this time. A desk top and drawer combination took up half of one wall and abutted a wardrobe with mirrored sliding doors. He had not noticed the other door in as he came in. This opened into a small shower and toilet facility which took up that corner of the cabin. Along the wall nearby was a long, concertina of pipes which were obviously an oil-filled heater. On the wall above there were racks for drying clothes. When he opened a small cupboard below the desk, he was delighted to find a small bar refrigerator which was stocked with several bottles of water and his very own favourite brand of whiskey. Several cans of mixers occupied the small racks on the door.

Breck had been too busy in appreciating the new facilities of his cabin to notice a large, cloth zippered bag placed upon the pillow of his bed. Curious, he pulled it over and pulled down the zipper. The bag opened out to reveal an astonishing set of cold-weather clothing. He laid each newly treasured item out onto the bed. There was a large, waterproof jacket with artificial fur-lined interior with a similarly lined wide hood, a thickly padded set of Arctic Rigger Coveralls, waterproof trousers, an alpaca woollen jumper, two thick flannel shirts, four pairs of thick woollen socks, two pairs of long woollen underwear and finally two sets of fur boot liners. Indeed, as he looked at the bags' contents it reminded him of his own personal Antarctic gear which he had packed for his last expedition. Everything had been provided including a pair of the latest snow goggles and face mask. Then he noticed a small printed note which stated that his snow boots were located within his wardrobe.

Walking over to the wardrobe, he slid opened its door to reveal a woollen white dressing gown, another larger waterproof jacket and below on the floor, a set of long snow boots. It took him a while but as he went through each item, starting with the boots and then each item of clothing, he found that they were all exactly his specific size. *Curious!*

He sat for a while on the small but comfortable chair at the desk and wondered about his new situation. He had been given the free time to go travelling due to the tsunami in Florida and had suddenly found an invitation to do some easy oceanographic work in the Arctic. Now, on board this small but apparently well-equipped ship, he found that his expedition needs had been prearranged down to the finest detail. It had not been asked for any of his personal details, yet everything fitted perfectly. *Even more curious!*

It seemed like only a short time before there was a knock at his cabin door. Breck got up and walked across the cabin and opened it. The tall First Officer stood a little back from the door, her friendly smile just as engaging as before. "Well, Dr. Breck. I hope that you have found the cabin to your liking?" she said with an air of confidence as if her question was purely rhetorical.

Beck returned the smile but with less confidence this time. He realised somehow that he was certainly not the master of his new fate. "The items of clothing…" he said quietly with some apprehension "… they are all in my size. Not an easy task of guesswork!"

The First Officer smiled again with a faint look condescension on her face "Oh we anticipated your acceptance of our invitation and did a little research. I hope that you don't mind." She said dismissing his curiosity with a wave of her hand in a casual gesture. "Now, let us go and see Captain Athanasios. He is really very interested in meeting you." She quickly turned and walked smartly back down the companionway, her long blonde ponytail swinging from side to side. Breck followed, finding that he had to walk faster than his normal pace to keep up with her.

Back in the main foyer of Deck 3, the First Officer pressed the 'up' button of the elevator and turned back to Breck and smiled. "We only have to go up one deck, but you may have noticed that there are no stairs leading there from this deck. Officially, there is no 'Deck 4' of the ship. Apart from the usual security concerns, we do not have such a deck due to Asian superstition as their word for four is 'Shi' and it is the word for death to them. We often have a number of Asian as well as

European guests aboard our other voyages and so we decided to respect our guests' superstitions."

The doors to the small elevator opened and the First Officer stepped in and used a key card to swipe a sensor which was set apart from the control panel near the door. Above this sensor was a small sign with red lettering which read 'Administration Only'. She pressed 'close' button, and the elevator ascended for a short time before they opened again. The foyer here was very small with subdued red lighting. The walls were generally bare of decoration with several narrow companionways leading off from them. It reminded Breck of the upper deck of some of the Navy ships on which he had served.

The First Officer turned to Breck and with her ubiquitous smile said in a soft voice, "Now that you will be a guest on board our ship, please call me Calypso for that is my first name and I am sure that we will be friends."

Beck was momentarily taken back by this personal request and with some uncertainty replied that she should also use his first name and then added "Calypso! Wasn't she the witch of Greek Mythology who kept the hero Odysseus a prisoner on her island?"

Calypso gave a short laugh and gave Breck an intense look, saying "Oh, you know your mythology then Terence! Perhaps you may have been a Roman playwright as well as an oceanographer?" she said, referring to the famous ancient Roman playwright of that name. She gently touched him on the arm and continued "Fear not! I am no witch. Not even to the crew like some First Officers, this is a happy ship and we all do our duty because we know what to do and love to do it. No, you will find that most of the crew of this ship, including its officers, mostly come from the Greek islands, and we value our ancient heritage. You will find that we even call our Captain Athanasios, 'Poseidon' because he is like a father to us and commander of the seas; very much like you in your navy may call your captain 'Skipper'. No, we are not gods here, but like the ancient deities of ancient Greek mythology we try to protect all of our natural environment, especially the sea. After all, there is only one true Creator."

They walked quietly along the companionway towards the bow of the ship for a short distance before coming out into a wider, triangular foyer. In front was a wide door which had the sign 'Bridge' marked upon it. To the left, on the Port side was another door labelled 'First Officer' and opposite that on the Starboard side was the Captain's cabin. She knocked quietly and a voice within called, "Come!". Calypso gave Breck one last smile and opened the door for him to enter. She followed him in and gently closed the door behind them.

The cabin was large and was again bathed in a subdued pale blue light. The walls were pattered with blue underwater scenes, and there was a large fish tank with a variety of marine fish within. In front of him, set at some distance back was a large desk at which the captain sat. If Breck had been impressed by the First Officer, he was now, a little in awe of the man who now stood up and walked around his desk to meet him. Breck thought that Calypso was tall, but her captain stood well above Breck's own six two. He was also proportionally broad, but his frame within the immaculate white uniform was also athletic. His skin had the same light olive brown colour, and his eyes were the same penetrating light blue of the First Officer, but his hair and full curling beard were totally white.

"Sir, may I present Dr. Terence Breck. Dr. Breck, this is our captain, Captain Athanasios" she said with some deference, smiled then quietly left the cabin.

The captain wore an immaculate white uniform of tailored slacks and high-buttoned jacket with a single row of bright brass buttons down its front. The ubiquitous trident symbol was prominent on each button and his epaulettes bore the intertwined four bars of a merchant marine captain. Walking around his desk with his right hand extended in welcome he said in a strong, but subdued voice without any trace of accent "Welcome to my ship, Dr. Breck. I am Captain Filip Athanasios, and I am very glad that you were able to accept our invitation to go cruising. Please have a seat." He indicated one of two armchairs which were against the wall opposite the large fish tank. "As you see, we like to be comfortable aboard the *Mystery* and I hope that you will also feel comfortable now that you are one of our special guests, – part of our crew in fact. Oh! and I hope that your family were not harmed by that small tsunami which hit the Florida coast the other day?"

Breck looked up, startled that the captain knew about that most recent event in his personal life. "How did you know about that? I only found out about it myself three days ago!"

"Do not be concerned, Dr. Breck. You should know that word gets around very quickly in our maritime world – especially strange events such as tsunamis in places where they are not common."

The captain pressed a small button on his desk as he pulled another comfortable chair over to near where Breck sat. Soon there was a small knock at the door and a orderly in a white uniform with a gold caduceus pinned to his left breast entered. The captain looked up and quietly said "Thank you, Damon. May we have some coffee?" he turned to Breck and continued "I am sure that you will find our Ellinikós kafés – our traditional Greek coffee much to your liking."

The orderly withdrew, softly replying, "At once, Poseidon."

The captain looked directly at Breck and said with some humour, "I hope that you will also get used to our informality about this ship. The crew like to call me after the Greek god of the sea, but there are no gods aboard this ship, just those who are here to protect the marine world of the Creator. The crew use that name for me in a manner very much like you may call the captain, 'The Old Man' in your own navy. Poseidon laughed then folded his arms, lent back in his chair and looked across his desk directly at Breck. "I have looked into your naval career also. You are a Lieutenant Commander in the US Naval Reserve I see, and regardless of your role as an oceanographer, you would have also done the usual basic training and taken the oath of defence of your Constitution. Your scientific work also suggests that you passionately care for the sea and can make independent judgements."

Breck was not sure where this conversation with the large captain was going, and he shifted uneasily in his chair. Poseidon continued: "This brings me to the reason for your invitation aboard my ship. Superficially, you are to be the Guest Oceanographer during this ecotourism trip into the Arctic Ocean. You will be expected to give the occasional talk to our other guests, conduct some relatively simple oceanographic, and perhaps perform some meteorological demonstrations with them – you may have noticed that we have most of the usual sampling equipment on our stern deck – and of course mingle with them socially just like the rest of our senior crew."

Breck thought that these tasks would be easy enough and nodded in agreement. The captain then leaned slightly closer to Breck in a confidential manner and, with a more serious look on his face continued: "However, that is not all. I hope that you will consider what I am about to say as being in the same realm as your Oath of Commissioning in your navy."

There was a long pause during which Breck remained silent; he now was very uncertain of his position aboard the ship. The captain stood up and half turned to a map of the North Atlantic and Arctic Ocean which was on the wall behind the desk. He turned and looked at Breck, "Your work on the bottom currents in the Antarctic was exactly what we needed for our voyage north along the

Norwegian coast into the Arctic Ocean. We wish to know exactly what is going on with the currents here and how they are affecting the North Polar icecap – what's left of it," he added. "But we need your help professionally…and for a while in secret. Are you interested? There is still time for you to leave our ship should you decline this position."

Chapter 6: A Change in the Climate

Breck was indeed interested in this mysterious proposal which the captain now proposed. He was by nature a curious man; it was inherent in his love of the sea, and of scientific investigation. He gave his word that, as a new member of the ship's crew, and as an officer in the US Navy, he would at least listen to the captain's proposal and, if everything was morally and legally justified, he would honour the captain's need for secrecy.

The orderly arrived with the strong, black Greek coffee, which was served in a small, copper pot which the captain called a 'Briki'. He poured the strong, sweet brown aromatic liquid into two white cups which had been provided. Breck found the coffee to be much to his taste. He and the captain enjoyed a moment of silence whilst they enjoyed their coffee.

Breck looked intently at the captain and raised his cup: "I think that I would like to join your ship, captain."

With a good-natured, "good!", the huge captain stood up and went to the cabin door and asked Breck to follow him. They walked back down the long companionway of the private Deck 4 to the foyer where the elevator was located. The captain pressed the 'down' button, and when the door opened, produced a small card key similar to the one which Breck had been given by the first Officer for his own cabin door.

As the door to the elevator closed, the captain also used his key card to swipe the sensor near the 'ADMINISTRATION ONLY,' sign but this time he held the card on the sensor until the elevator began to descend. He turned to Breck, holding up his card. "Your card will do exactly what mine will do. Hold it on the sensor for extra time and you will be able to descend, but you must only do this if you are alone. The other passengers are not part of our private research. Do you agree to this rule?" he said sternly.

Breck agreed to this and stood back from the elevator's doors for some reason. Perhaps he was now beginning to think that he had taken on something beyond his normal expectation. He was even more apprehensive as he noticed that the elevator continued to descend past the "Deck 1" level - the lowest level on the indicator board.

The elevator's doors opened into a narrow companionway illuminated by shielded red lamps. The door opposite, Breck suddenly noticed, was a typical robust watertight door typical of those with which he was familiar in naval vessels; solid with rounded corners and with several securing 'dogs' or clamp handles around three of its sides with a raised splash guard at its base. Indeed, as he followed the huge captain along the companionway he felt as though he had suddenly been transported to a 'below decks' section of one of the many US Navy ships on which he had served as an oceanographer. However, here there was none of the usual 'hustle and bustle' typical of a crowded naval vessel with crew going about their duties: instead, there was only silence and the lonely confines of the red lit companionway with its watertight doors on either side. There were no portholes nor any other illumination other than the shielded red lamps which clung to the low deckhead above like some upside down glowing red beetles. Finally, they reached the end of the companionway, where the captain opened another watertight door which led into a large chamber.

Breck was astounded at the size of this chamber which was at least two decks in height and seemed to take up the entire width of the ship. It was more like a small aircraft hangar with a flat roof festooned with small cranes and other heavy lifting machinery. When he looked down at the deck, he was amazed to see a large, open pool of water, gently moving back and forth with the gentle motion of the sea outside and glistening black in the red light of the lamps positioned vertically around the walls of the immense cabin. He had only seen such a Moon Pool once in his career, for this is what his research colleagues had called such a feature when he was a guest on board a deep-sea diving rig. Such a pool had been used on that rig as a place of easy retrieval of deep-sea diving equipment including small submersibles. He also remembered being told with some open jocularity by a senior naval officer that the CIA often contracted such vessels for use with their covert ocean activities under the pretence that their ship was engaged in scientific sea-floor research. Certainly, there had also been no mention of this ship's additional oceanographic feature deep within its hull in the glossy magazine extolling the ship's facilities.

Captain Athanasios saw Breck's astonishment and, with a confident smile said, "Well now, Dr. Breck, what do you think of our underwater facility? We anticipated your acceptance and so we have acquired all of the more specialised oceanographic equipment and instrumentation which you might

need for your – shall we say, 'additional research'
- with us."

Breck could only look around in wonderment. There was indeed a comprehensive range of oceanographic equipment: several recent variants of Box Corers; CTD Samplers[3]; and a large range of other sensors hanging from small cranes attached to the deckhead above the pool. There were various lockers and racks along the other three walls of the cabin which were labelled with smaller instrumentation needed for deep ocean investigation. Inside the pool on the far side, Breck could just make out the glass dome of a small submersible vehicle. There seemed to be duplicates of all of the instruments which he had used in his own research in Antarctic waters, and several other items which he recognised would have been useful in those past days of his research if he had had the funding he had wanted.

---

[3] Conductivity – Temperature – Depth sensors

Breck followed the captain around the edge of the large pool and through an open watertight door which bore the label, "MONITORING STATION". Inside, two members of the crew jumped to attention at their station. They were also tall, but unlike the other crew members whom Breck had seen, these two men had pale, almost white skin which seemed red in the red glow of the muted night vision lamps. Their hair was black, accentuated even more in the red light. Poseidon waved his hand in greeting and introduced Breck, giving their names as Fionn and Rónán, then softly told them to get back to work. Poseidon turned to Breck and explained that these two men were from the northern isles of Scotland and were the ship's specialists in cold water diving. They had also been trained thoroughly on the use of the ship's oceanographic equipment and were there to help Breck in his work. Breck gave the men a nod of his head in acknowledgement, and then looked slowly around this new cabin. It reminded him of his own laboratory on board his previous oceanographic vessel in Antarctica. There was a long, white bench along the far wall which contained a variety of computer screens and computer terminals, including a large monitor for SIMRAD brand Side-scan Radar (SSR) which was now set on 'standby'. Everything which he needed, and perhaps more, for examining the ocean floor and the sea's different levels of composition.

The big captain turned to Breck and waved his hand across the view of the cabin. "You see here Dr. Breck, the latest technology for looking under the sea, and a willing crew who know how to use it. But then, I see from the look on your face that you know that already."

"I am most impressed Captain Athanasios. I wish that all of my previous expeditions were as well equipped as your ship. How do you see me using such equipment for simple bottom current measurements?"

The captain smiled and turned back towards the door. "Oh, as you can see you won't be alone. Some of my people such as Fionn and Rónán here, have had some specialist training on these instruments. You will meet them more professionally once we have gotten underway, and then you will become more acquainted with my staff and our methods of operation. But come! We will return to my cabin, and I will explain in more detail why we have invited you on board the *Mystery of the Seas*." With that, the captain went back out into the Moon Pool cabin, and into the long companionway which led to the elevator to the upper decks. Breck felt even more like a small boy in a magical chocolate factory. He was also a little confused; *what had he gotten himself into!* he thought.

Back in the captain's cabin, Breck was again seated facing the huge map which was on the wall behind the captain's desk. It showed the North Atlantic with a track drawn upon it which travelled from Edinburgh over to the Norwegian coast then along it and into the Arctic Ocean. The captain stood next to the map and ran his huge hand along the track and, turning to Breck, said "This is our proposed route. We will be sampling the surface and bottom currents along the way, both Underway and On Station." He smiled and gave a short cough then continued, "of course, when we stop for any critical On Station sampling you can be showing our guests how it's done with some of our more simplified sampling techniques, Nansen Bottles[4] and the like for example, on the stern whilst our team will be conducting the more sophisticated sampling below through the Moon Pool – after your prearranged instruction, of course."

Breck had little to say but saw the need for deception in the captain's method. Conventional oceanographic sampling along a country's coastline would attract considerable maritime attention, especially once they went into the Arctic Ocean and closer to the Finland – Russian border. The captain sat down and looked across the table with a more serious expression on his big face. "You have an expression in your country.

---

[4] Nansen Bottles - The Nansen Bottle is designed for the capture of water deep in the ocean. It was designed in 1894 by Fridtjof Nansen, the Norwegian explorer and scientist.

Something about teaching your grandmother to suck eggs or some such. So now I hope that what I am about to tell you is not too obvious for a scientist of your training and experience. Let me explain the reason behind our mission and why we wanted you to be part of it."

The captain stood up and pointed to the centre of the North Atlantic on the map, "As you well know, our oceans have surface currents which then sink and return as sea floor bottom or counter currents as part of the thermohaline circulation pattern. This circulation is like a giant caterpillar tread, with warm currents flowing across the ocean's surface to colder latitudes where they cool, become saltier and denser, and then sink to the sea floor. Here they flow back again in the opposite direction as a counter current. Such a system is that of the Gulf Stream which is the warm and swift surface current originating in the Gulf of Mexico, and which then flows up the eastern coastline of the United States. It then veers east off the coast of North Carolina and moves across the Atlantic towards north western Europe as the North Atlantic Current. This why the coasts of northern USA and Canada have such cold climates whereas, way to the north, Scotland and northern Europe experience much milder climates than those away from the current. At its northern extremity, just south of Greenland but possibly not far from the Arctic Ocean, it forms the North Atlantic Deep Water which then sinks and moves south as its counter current. This is part of what is called the Atlantic Meridional Overturning Circulation or AMOC. It is this North Atlantic Current driven by its thermohaline circulation which is the main object of our research."

The captain returned to his seat and sat down, looking intently across his desk at Breck as he continued. "Recently, two Danish researchers, Larsen and Bjornsen, have published their findings from their research along the Norwegian coast. Their study has suggested that this circulation is slowing down and in fact may soon cease. They think that this reduction in flow rate is due to the continued melting of the Greenland and Arctic ice caps due to global warming and the increased introduction of fresh water into the system. The implications of such a disruption to this circulation may have a very disastrous effect on the world's climate."

The captain stood up again and walked around his desk until he stood near to where Breck was sitting. "Imagine what would happen if the warm Gulf Stream ceased and the cold counter current did not flow south! All of coastal North America on its east coast and Europe on its north and western coasts would suddenly experience freezing temperatures much like that of eastern Canada."

Breck looked up at the captain and saw the look of concern on his broad face. Breck was not too sure as to what to say about this startling revelation, but he replied quietly, "Well! I hope that these findings would startle the UN climate change people into further research."

"Alas, Dr. Breck. The UN are not convinced. They are sceptical about this one paper which has been published and its dire hypothesis. They want more evidence and they, like most of the world's population, do not want to really think about any catastrophic consequences. But tell me Dr. Breck, what thoughts do you have about such a reduction in bottom currents based on your work in Antarctica?"

"I am not certain that my work in Antarctica would be able to support the Danish scientist's findings. It was a bit more complicated than the general picture which you have painted. We sampled the sub-surface currents along the warm Brazil Current which flows southwards along the Argentinian coast to off the Rio de la Plata at about 38^0 South latitude. Here this current meets the cold northward flowing Malvinas Current and this convergence, called the Brazil-Malvinas Confluence, has a complex pattern of both warm and cold rings and eddies. Whilst we sampled this current right down into the Weddell Sea in

Antarctica, we found the results to be rather inconclusive. However, it is thought that the southward flow of warm water of the Brazil Current may have some influence in the overall action of the AMOC further north in the Atlantic. Indeed, some of my former colleagues even suggested that the mass melting of some of Antarctica's ice would introduce more cold, freshwater into northern waters, slowing bottom currents dramatically as these freshwater currents would not sink but stay on the surface. Some even suggested that such a reduction in the overturning of currents would stop the sinking of dissolved carbon dioxide in the warmer water and so leave more of this gas in the atmosphere thus increasing global warming rather than stopping it."

The captain returned to his desk and moved his large hand across the expanse of the Arctic Ocean and northern Russia. "Unfortunately, there is more to the problem than just the reduction of ocean currents. As you are well aware, increased global warming has melted a significant part of the ice cap over the Arctic Ocean to the point where it is estimated that by 2040, the ocean there will be completely free of ice. A good thing, perhaps for those who are now using the Arctic Ocean as a new sea route between Russia and Asia, but not good for our world's climate." The captain sat down and continued: "Perhaps you may remember the theory of Ewing and Donn who put forward their hypothesis about how the great ice ages were formed?"

Breck smiled "Yes well, we did cover that briefly in my undergraduate years. I think that they proposed that the various ice ages occurred at times when the Arctic ice cap had melted, but their ideas where generally disregarded because they could not explain what would cause this melting?"

"Yes, true enough. But that was back in the 1950's when the theories of plate tectonics and global warming had not yet come upon the scene. Never-the-less, their hypothesis now has some relevance. If the Arctic Ocean is free of ice, they argued, then there would be mass evaporation of surface water from it. This water vapour would not have to travel too far, especially on the fierce Arctic winds which would blow these huge masses of water vapour onto the colder land surface of northern Russia and Canada. Here it would condense and fall as extensive snowfalls. With successive and continuous snowfall due to this evaporation, large volumes of snow would be compacted on the land to form ice which would eventually build up and once again flow inland as large ice sheets; a new great ice age!"

Breck was stunned by this explanation by the captain. He was a scientist and would need to know more data than that given in just one research paper, and the recycling of an old, discredited theory. But the captain continued as though he had read Breck's thoughts: "We will need more research before any of this can be made public. That is why we will have both an open and closed set of oceanographic measurements on the *Mystery*. As far as the ice age theory goes, we will be having another guest coming on board at Bergen in Norway who is well into that line of research. But then I know from my people who watched your teleconference in Edinburgh, that you are already acquainted with your future colleague, Dr. Yuri Ziminov from the Siberian Academy of Natural Science in Krasnoyarsk."

Breck remembered the Question Time at the end of his talk at the teleconference and how a Russian participant – he had already forgotten his name – had asked some pointed questions via the video link from Russia. Breck had not thought much about his video meeting with Ziminov as much of the time had been taken up by the Russian talking about his own research which had little to do with Breck's talk. The captain continued: "We will be meeting Dr. Ziminov in our first port of call which is Bergen. He has been able to get permission to leave Russia and come on this expedition of ours simply as a guest speaker, like yourself, but on his subject of climatology. I am sure that after what I have told you, the two of you will have a lot more to discuss. I might warn you, however that Dr. Ziminov's hypothesis and work has not gone down well with some of his compatriots. Indeed, in his brief communication to us in accepting his appointment, he suggested that there are those in Russia who would much prefer that he did not continue his research and that he felt threatened by them."

Breck remembered now that his brief encounter with Dr. Ziminov during the conference had been one of urgency on the Russian's part. He seemed to be trying to get his ideas across almost like a drowning man calling for help. These new revelations worried Breck but he was still intrigued by voyage yet to come. "So then, captain, how will this affect my part in this voyage?"

The big captain smiled. "Do not be concerned with any possible trouble from the Russians. My organisation is well able to handle any interference to our quests. To put it in simple terms, we will endeavour to sample the currents along the

Norwegian coast and perhaps into the Arctic Ocean, and if the results confirm what our two Danish researchers have suggested, then we will ask you to publish these results as a paper in the usual reputable international scientific journals."

"How do you propose to do that? No offense, but after all, few scientific bodies and certainly not the UN, would accept data from a private ecotourism voyage."

"Naturally they would not! However, if we confirm that there is a weakening of the currents, especially the Gulf Stream and if Dr. Ziminov's data seems creditable enough, then my organisation will provide your Naval Oceanographic Office with unlimited funding for you to research these currents in detail, and also to assist Dr. Ziminov's work and only then could you, and perhaps Dr. Ziminov, publish your findings internationally. Your international reputation, and that of NAVOCEANO, would be more than enough to sway scientific opinion to bring global action to prevent any future catastrophe."

Breck stood up and walked across the cabin before turning back to where the captain sat. He was still very much troubled by what he had heard. "Thank you, captain. That seems practical enough, but may I ask you, what is your organisation, and how does it get its funding?"

The captain stood up and leaned across his desk. With a faint grin on his broad face, he said with a slight hint of humour in his voice: "Ha! I thought that you would be curious to ask that question and rightfully so! Well, there is nothing sinister about us. We simply refer to ourselves as 'The Organisation,' as we have no need for pretentious names to appeal to the general public nor ask for funding by selling soft toys and the like. In fact, we go to great lengths to conceal our activities so that we do not attract adverse criticism nor physical opposition, and so far, we have been most successful for a very long time. As to funding, well we have the oceans of the world to thank for that. Humankind has long been careless about the many treasures which it has lost to the depths, and we are very good at salvage. No! Be assured that when I say that you will have unlimited funding, I mean just that. Do you have any more questions, Dr Breck?"

Breck was still stunned with the magnitude of the cruise of the *Mystery of the Seas* and its possible outcome. If the captain was right about the future of the northern currents and of the sudden change in the world's climate, then it would be very difficult for him to withdraw from the project. He would need to think more clearly about the fine detail, so he replied: "No! I have no questions for now Captain Athanasios, but I will have to think more about what you have said."

"Good! I see that you have a lot of information of great importance thrust upon you, so I will always be available for any questions which you may have later. I will have my orderly take you back to your cabin as we should be sailing as soon as the last of our guests have come aboard. I will ask First Officer Thalassa to escort you and our other guest speaker, who is currently on board, to our initial 'Captain's Greeting' after we sail – say at about 1700 hours? Now I am sure that you would like to return to your cabin and take in some of the details which I have laid heavily upon your shoulders." With that the captain stood up and walked with Breck to his cabin door, which he opened to reveal another tall man in the white uniform of a steward waiting. "Until five, Dr Breck."

Chapter 7: Unwanted Guests

Breck returned to his cabin and pondered on the revelations made clear by the captain. Certainly the science had sounded convincing, and he was equally certain that Captain Athanasios was serious about the ship's mission. He was also concerned about his own tasks on board the *Mystery of the Seas,* now that he knew exactly what was expected of him. Being a guest scientist on board an ecotourism cruise ship was no problem for him, but being involved in covert operations planned by a mysterious environmental group which simply called itself 'the organisation,' raised some concern. *But it was too late now*! he thought. The objective of the cruise was important enough for him to remain and assist in its finalisation. He was in the position to carry out his shipboard tasks, both covert and overt, and he certainly would not miss the opportunity to carry out his own personal research after the cruise, especially with the promised unlimited funding.

He resigned himself to his fate and got on with the mundane tasks of putting his personal belongings into the wardrobe and drawers provided in the cabin. He had found that in his absence, more items of clothing had been added to his wardrobe. He went over all of the new items, and marvelled at the quality of the garments and footwear which seemed to have been made to order. There was even a uniform as such consisting of black trousers, highly polished black shoes and a white naval shirt with the ships name and trident logo above the breast pocket. He also found a white dinner jacket, two white shirts and a black bow tie in the other side of the wardrobe. There was a small shudder of the ship and noises from outside which told him that the *Mystery of the Seas* had moved away from the dock and was heading out to sea.

Breck showered and put on the uniform which had been provided, white slacks and shirt with plain epaulettes and the ubiquitous trident symbol on the left breast. A perfect fit, and there was also a name tag with his name and the subtitle *Dr. Terence*

Breck, Oceanographer. He quickly changed out of his civilian clothing and looked in the long mirror of his wardrobe. Not quite the look of authority as his uniform as a Lieutenant Commander in the United States Navy, but it would help him play his part as a senior member of the ship's crew. He had just sat down at his desk to make some notes as to his planned course of lectures on oceanography when there was a knock at the door. Opening it, he found the lovely First Officer waiting outside.

"Come along, now Dr. Breck. We have to join our other guest speaker and the captain on the stage in our small theatre." She said formally with just the hint of a mischievous smile as she extended her arm out ready to accept his. Breck looked around the cabin in case there was something he should take along with him – a clipboard perhaps. He had long been induced to always carry a clipboard when walking around naval establishments. He had been told during his first induction into the navy that a clipboard was not only useful in carrying short lists of notes, but also to give an air of importance, which would display some confidence in the performance of his duties. There being no such false crutch upon which he could lean, he took the First Officer's arm and returned her smile, albeit rather limply being uncertain about what was to follow.

They had only walked down the companionway a short distance when the First Officer let go of Breck's arm and knocked on another cabin door. There was a muffled sound within and a short delay until the door opened revealing a middle-aged man of average height with unkept red hair and bushy eyebrows to match. He had also changed into his uniform but on him the shirt did not seem to fit comfortably over an ample belly. He looked up at the tall officer and mumbled something in a rich Scottish brogue as he came out and closed the door: "Och! Sorry fur th' wee delay bit ah was huvin troubl' wi' th' fancy uniform."

The First Officer smiled and stood back, looking at Breck. "Dr Terence Breck, Oceanographer – allow me to introduce Dr. Hamish Fraser, Marine Biologist."

Breck stepped forward and put out his hand which was taken by his new companion. "Glad to meet you, Dr. Fraser. I'm glad that we have a Scot on board. I hope that we will have time for you to explain some of the culture of my ancestor who left Scotland many years ago."

Fraser looked at Breck and raised one bushy eyebrow: "Och! an American then!" he said with a smile, "weel that is guid. We can blether aboot oor countries whin we git tired o' all o' th' scientific stuff."

Breck laughed, "Then I'll look forward to some chatter and a few drinks then after work."

"Aye! a few wee drams would be guid right at this moment though. Ah'am no' guid in th' public eye ye ken?"

Breck liked his new companion and was just about to suggest that they find the bar to seek some additional support when the tall First Officer quickly took them both by their arms to walk with her along the narrow companionway. "Now gentlemen. There will be time for that after you have been introduced. The captain and our assembled guests are waiting"

They walked along the deserted companionway and down the stairs towards the stern of the ship. They passed a set of plush double doors, rounded a corner and continued on to what Breck imagined would be the ship's version of the 'Stage Door'. He had been correct, and saw that there were a group of uniformed ship's crew already waiting in the wings for their guests out front to be seated. Captain Athanasios stood near the side of the stage hidden from the audience by a long, blue stage curtain. He turned and smiled at the three new arrivals, then turned back and confidently walked out onto the small stage. In the audience, those who were standing immediately sat down. Talking ceased as all attention was focussed on the imposing figure who now stood smiling before them. Breck noted that the auditorium, if he could use such a grandiose term, was really just a large room which contained just enough seats for the small number of passengers. The stage itself was only about one step high but as the floor of the room sloped down towards it, there was more than adequate exposure to see those standing upon it.

The captain walked up to the very front of the stage - perhaps, no more than an arm's distance from the first row of passengers - and smiled. Without the need for any microphone, he said in a friendly voice: "Welcome aboard my ship the *Mystery of the Seas*. I am Captain Filip Athanasios, and it is my pleasure to take you all on a voyage of discovery to seek out some of the sea's mysteries." There was a small amount of supressed laughter at the captain's play on the ship's name. Without further ado, he half turned and said. "Now it is my additional pleasure to introduce senior members of my crew. Firstly, I would like to introduce a very capable lady who really runs this ship without whom I would be 'all at sea' – First Officer Thalassa."

The tall First Officer walked confidently onto the stage and stood at the right hand of the captain, giving a very big smile to the captain and a brief wave of her hand. There was applause from the audience and a few quite gasps from those who appreciated her statuesque beauty and rank. She returned to where Breck and Fraser were standing and gave a quick knowing smile and then turned straight-face towards the audience.

Whilst the captain continued to introduce the senior members of his crew, she edged closer to Breck and without turning away from looking at the audience said quietly:

"Do not look at me but what do you notice about our audience, Lieutenant Commander?"

Breck was taken aback by her use of his naval rank and quickly looked more closely at the assembled group of people sitting in the room. He had been on several cruises with June out of Fort Lauderdale and at first glance the people in front of him seemed to resemble the usual crowd: mostly middle aged to elderly, probably well-healed by the look of their clothing and mostly couples or prosperous looking singles. He saw nothing much out of the ordinary for such a cruise ship. The First Officer continued quietly:

"Yes, most are your usual cruise passengers but remember many are here by invitation, politicians, senior media people and well-known creditable environmentalists. Some have also paid a lot of money to come on our specialist cruise. Do not look in their direction but in the far right of the room, sitting almost by themselves are two men who appear slightly younger and more solidly built than our other older passengers, the two men in the shabby grey suits. We spotted them as soon as they came on board. They say that they are from some Ukrainian environmental agency, but Captain Athanasios believes that they are either current or former SFB[5] members. The short one calls himself Danylo Kavalenko and the other is Yuri Pavluk. The captain has assigned one of our Senior Stewards as their personal cabin steward. Armenus is his name, and he has sailed many times into the Black Sea and is a very clever fellow with skills in many languages. He easily played the unobtrusive cabin steward, and he has found that whilst he was tidying up their cabin when we were leaving port, that his two guests spoke entirely in Russian. We must keep a close eye on those two, but I am sure that you can play the game of secrecy like the rest of us."

[5] SFB: Russian Federal Security Service – the successor to the KGB.

Chapter 8: Bergen

Breck had never been to Bergen before. The ship had come into the Vågen, the long, narrow central harbour of the city, sometime in the early morning and so the ship was unusually still and calm when he awoke. This had been a welcome relief from the rough sea conditions which they had experienced crossing the North Sea over the last two days. They had turned at the island of Fedje at the mouth of the Hjeltefjorden, that long fjord which runs southeast into the approaches to the port of Bergen, sometime in the early morning and had tied up at the old Hanseatic League wharf at Bryggen, the name itself simply meaning, 'The Dock' in Norwegian.

Standing up on his bunk, he looked out of the small porthole of his cabin and saw the long line of brightly coloured buildings which lined the old wharf. He had read that the wharf itself had been built in the 14th century for the Hanseatic League, that vast trading group of northern European commercial guilds, but these beautiful wooden buildings were only of the most recent construction after the many fires which had swept the area in the past. Most of the buildings consisted of three and four stories, usually with a central door and two windows on the ground floor and a garret window below a steep, gabled roof. The buildings were brightly painted in shades of red and yellow as well as white, they were all built against each other to keep out the cold.

He showered, dressed casually in slacks and a plain woollen shirt and went up to breakfast as usual. The dining room was still mostly empty, and he guessed that most of his fellow passengers were still in their cabins getting over the ship's recent rough passage. His only compensation to that was to have very strong black coffee instead of his usual Cappuccino. Scrambled eggs on toast and several rashers of bacon helped.

He retired to the ship's small lounge bar to see if there was also a morning paper but found none.

Perhaps he would find an English-language edition in Bergen which he planned to visit after the mad rush of exiting tourists which he assumed would happen later that morning. He was surprised when First Officer Thalassa came into the room. Breck was stunned by her beauty and attire. She was no longer in her official, no-nonsense sea uniform but dressed in white slacks and a woollen, roll-neck jumper also in pure white but with a wide, blue band consisting of a continuous line of internally folding squares, as a repeated motif. Her long blonde hair was tied up into a long ponytail which projected out from the back of a white baseball cap which she wore. With her very tall and well-proportioned body, Breck thought that she looked every inch the ideal of a modern Viking Shield Maiden.

Calypso smiled when she saw Breck sitting along and strode over to his table. "Come now, Dr. Breck! It is going to be a nice sunny day here in Bergen. Time to get up and walk though this beautiful city." She said extending her hand down to help

Breck get out of his chair. "Come! We can spend some time at the Fish Market before seeking out our new Russian colleague, Dr. Ziminov at his hotel which is nearby."

Breck met her enthusiasm with an equal response.: "Thanks. That would be great, but I had hoped to avoid my fellow tourists."

"Oh! Don't worry about them. They will probably not see the light of day until the, 'Sun comes over the yardarm,' as the old sailors used to say. And don't worry about visiting the Fish Market! We are not here to buy fish and besides, this market is outside with many colourful stalls selling other foods and craft items as well as seafood. It has become very trendy with tourists, and it will be open soon. Come! We must beat the crowds."

Breck took her proffered hand and stood up, put on his jacket which he had discarded over the back of his chair before following her out of the lounge.

The security gate and gangway were already set up and Calypso gave the two security staff a friendly wave of her hand as she walked right through the gate without stopping. Breck followed close behind, unsure of himself and what he should have presented at the security gate. The two staff members simply smiled as he passed through and walked down the angled gangway to the paved wharf below.

It was indeed a lovely morning making Breck feel uplifted as he walked along the pathway which separated the blue, sparkling water of the Vågen from the road which ran along its side. Across this road were the many old wooden three-story houses of varying bright colours and arched gables. Soon these buildings changed into more substantial stone and brick buildings; some of them built with mock medieval casements but still with modern shops on their ground floors.

The end of the path curved around at the end of the waterway, suddenly Breck found himself facing one of the main thoroughfares of the modern city of Bergen with all of its traffic and noise. A short walk along the pathway at the end of the Vågen brought them to the Fisketorget or Fish Market. This consisted of numerous interconnected stalls sheltering under maroon-coloured tents which also had transparent plastic sheeting sides. Breck was astounded at the great variety of fresh sea food, fruit and vegetables but also the wide variety of items including local crafts and hot food for sale.

Calypso allowed Breck to browse some of the stalls before quietly saying: "Too many dead fish! I much prefer the living ones in the oceans. Let us move on to the hotel and pick up your new colleague. Then I will take you both to a nice coffee shop for some real Norwegian treats." Breck followed her out of the markets and as she turned once more to walk along the western side of the narrow harbour. A broad walkway along the edge of the waterway led passed the many expensive-looking white yachts and motorboats which were moored there then back onto the main road, which ran north parallel to the harbour. On the other side of this road was a line of hotels; mostly of six or seven stories with shops on either side of their wide entrances. Calypso crossed the road and went into one of the double door entrances above which was the sign 'Karlsgaarden Hotel'. Beck followed and caught up with her in the main foyer.

Despite the conservative nature of its exterior, the hotel's foyer was both modern and expensive looking. There was a curved reception counter at one side richly decorated in separate wood panels, behind which stood a young lady in a smart blue jacket. Calypso walked confidently up to the counter and said in flawless Norwegian: "God morgen. Vi vil gjerne se Dr. Ziminov. Jeg tror at han bor på dette hotellet."[6]

The young lady's welcoming smile suddenly vanished and the pen which she was holding dropped out of her hand. Her face had turned pale

---

[6] "Good morning. We would like to see Dr. Ziminov please. I believe that he is staying at this hotel."

and she stammered in reply: "Åh! Åh! Vennligst vent. Jeg skal hente manageren."[7] And quickly ran into an office at the rear of the counter. In a very short time, an elderly man in a dark blue suit returned with the young lady following close behind. Calypso looked around the manager and said to the young lady in English: "Is there anything wrong?"

The manager came right up to his side of the counter and said in a low voice as if not to be heard past the desk:" I am sorry." He said in very good English with little trace of an accent "but Dr. Ziminov died in a tragic accident two days ago. His body has only just been taken to the morgue and the police are trying to contact his relatives." Calypso turned to Breck with little emotion in her lovely face then turned back to the manager. "How did he die?" she asked.

The manager opened his arms wide in despair and said: "We had a fireworks display in the square

---

[7] "Oh! Oh! Please wait. I will get the manager."

along from the harbour and the police think that Dr. Ziminov must have leaned too far over his balcony to view them and fell. It was four stories, you understand, and he was killed instantly. A tragic accident." There was a long pause when the manager hoped that his explanation would be enough. He had had more questions from the police than he cared to answer the previous day. "Are you from the ship by any chance? Dr. Ziminov told me that he expected to join a ship today."

Calypso answered that indeed they were and that she was the First Officer come to take him to it. She also produced a business card identifying her and gave it to the manager. He looked down at the card and gave a short grunt of satisfaction then looked around Calypso at Breck and said: "Would you happen to be Dr. Breck sir?" Breck stepped up to the counter and replied that he was indeed Dr. Terence Breck, a guest speaker on board the *Mystery of the Seas.*

There was an expression of relief on the manager's face; one that suggested that now he had some form of closure following the unfortunate death of the Russian scientist. He quickly bent down below the counter and came up with a small package wrapped in brown paper. He said with some solemnity: "Dr Ziminov left this here for you, sir. He seemed to be rather agitated at the time but wanted you to have it if he was not able to give it to you in person."

Breck took the curious package and looked at Calypso with some apprehension. She quickly turned to the manager and asked if there was anything which they could do, explaining that her business card had the appropriate contact details should there be any enquiries from the police. The manager replied that Dr. Ziminov had paid for his room in advance and was looking forward to leaving on the ship. Again, the manager suggested that his recently-departed guest had had a feeling of foreboding, but at the time he had simply put it down to the uncertain Russian nature of being in a foreign city. Thanking the manager for his assistance, Calypso looked at Breck then turned and walked slowly out of the hotel and headed back down towards the Fish Market and the ship.

They had only walked a short distance when Calypso said in a quiet voice without turning her head: "Don't look now Terence, but I think that we are being followed. Let us stop at the next shop and pretend that we are looking at their goods and you may be able to see who is following us."

They stopped at a small shopfront and looked into the glass to observe the window display. Looking slightly back the way that they came, Breck saw two men who also had stopped and were looking nonchalantly out at the harbour. One was a big man who had the physique of a prize-fighter and the other was a small, thin man; the brawn and brains of the outfit.

"Let us walk on and turn into the next street to see if they follow us there." Calypso said hurrying on.

Breck followed her as they turned into a narrow street which came up between two of the hotels. It was a narrow street; no more than one vehicle lane and to her dismay, Calypso saw that it ended in a service entry to the rear of one of the hotels. She ran up to the end of the street and saw that the roller door of the service entrance was not only closed but also padlocked. Breck had followed her closely and when they both turned, they found that the two men had also come into the street and were advancing slowly towards them. The thin, short man stopped but his large colleague continued and pulled a switch-blade knife from his shabby grey coat pocket.

Breck now wished that his Navy training those many years ago had also include unarmed combat. No such luck! Never-the-less, his instinct was to move in front of his female companion as their adversaries came closer.

"We want the papers which Ziminov left for you. Give it to me, now!" the larger thug said in heavily accented English, thrusting is open hand out towards Breck. Before Breck had a chance to reply, Calypso suddenly took one step forward with her left foot moving quickly past Breck and the other shot up in a rapid, straight-legged kick so that the heel of her sturdy walking shoe caught the thug just under his jaw. The big man was physically lifted up and his large body was shot backwards striking the wall of the building before collapsing in an unconscious heap. Breck was astounded at the speed and force of this defence, but he was even more surprised when Calypso took two more steps bringing her at arm's length to the smaller man. She grabbed his throat in one hand and lifted him bodily off the ground and shook him like a fox shaking a rabbit. She then threw him hard over against the wall where he fell unconscious on top of his big colleague.

She turned to Breck, shrugged her shoulders and gave a depreciating smile, "He was only a little fish. Time to get back to the ship and report to the captain." With that she stepped past the crumpled bodies of their two assailants and walked back down the narrow street to the main thoroughfare and the harbour. Breck recovered his shock and quickly followed.

Chapter 9: The Attack

Breck had to walk fast to keep up with the tall First Officer on their way back to the ship. There was little conversation; the events of the morning now required some urgency and few words. Calypso brushed off Breck's admiration for her fighting skills by a simple laugh and the reply that as a young girl growing up around the docks of Gozo in Malta, she had to learn to fight. She also added that she had trained in Japanese Mixed Martial Arts and had reached the high rank of First Kyū[8].

They returned to the ship and hurried past Security and straight to the captain's cabin. Luckily Poseidon was still there, finalising some of the necessary paperwork needed for their extended voyage north along the Norwegian coast and into the Arctic Ocean. With little formality, Calypso gave a detailed account of the death of Dr. Ziminov and their encounter with the two thugs; omitting her defensive actions and simply saying that they

---

[8] First Kyū is the highest grade of Brown Belt in many MMA schools.

were able to escape uninjured, despite being cornered. Poseidon bade them both to sit down whilst he pondered the next course of action.

"It seems that the Russians are very serious about stopping our voyage. Dr. Ziminov's work would have been vital to our objectives. You say, Calypso that the late doctor had a package?" Calypso turned and looked at Breck who had been carrying the mysterious package which had been given to him by the hotel receptionist. He had been carrying it under his arm in a firm grip as though his life had depended upon it since leaving the hotel. He placed it now in front of the captain who pulled a small knife from his desk drawer and slit open the package seals. It contained a large, spiral-bound notebook, about A4 size and with the title "Доктор наук Юрий Зиминов, научный сотрудник[9] "

"It looks like Dr. Ziminov's research notes. This should help us in our work, but I don't understand Russian." Breck said.

---

[9] Doctor of Science Yuri Ziminov, research fellow

"That will be no problem, Dr. Breck. I will have our steward Armenus come to your cabin once he has finished shadowing our two "Ukrainian" guests for the day. He is very fluent in speaking, reading and writing that language, but he will need your help with the scientific aspects of the document," Poseidon said handing the large notebook back to Breck. "I think that, considering these new circumstances, that we will have to keep a closer eye on our two guests. It appears that the Russians are quite prepared to kill those who may stand in their way. We will all need to continue behaving as normal but take extra care when dealing with our Russian guests."

Breck left the captain's cabin somewhat concerned. He had signed on as a guest speaker in his much-loved subject of oceanography, and now he was involved in a deadly game between the captain and crew of the even more mysterious ship and some as yet unknown Russian group. A potential colleague had been murdered and he had been involved in an attempt on the life of the ship's First

Officer and himself. Back in his cabin where he took the insecure step of locking its door, he sat on his bunk and thought about the situation. Poseidon had been right, of course. There was nothing more to do but continue doing what was needed, but with extreme caution; he would have to play the game as required and not show the two Russian passengers that their true motives were known.

So Breck put his apprehensions aside for a moment and got on with preparing the more practical side of his shipboard tasks; both as a guest speaker and as a clandestine oceanographer. Nothing much had been said about the unfortunate death of Dr. Ziminov except that Captain Athanasios had announced at dinner on the first night out of Bergen that their resident climatologist had been called away at the last moment and would not be joining the cruise.

And so, the *Mystery of the Seas* began her slow cruise north along the coast of Norway. The weather had been remarkably fine for that time of year and the crew provided the ship's guests with a wide variety of daytime activities and nightly lectures on a variety of subjects dealing with the sea. Dr. Hamish Fraser, the marine biologist had given an introductory talk on the kinds of animals which could be encountered in this part of the world and Beck, who had already given several small talks on the development of the science of oceanography as they crossed the North Sea, now began his series on the measurement of the ocean's environment. The ship's Internet had given him all of the notes and photographs needed for his presentations and the guests had all responded well to his account of how the early records kept by seafarers had developed into a modern science in the Nineteenth Century. He now prepared a comprehensive set of talks about the general methods of oceanography, including its methods of sampling, including "Under Way" whilst the ship was moving and also sampling done "On Station" when the ship needed to stop at a precise location and carry out tests whilst stationary. He did not refer to the true scientific capabilities of the *Mystery of the Seas.*

On the second day out from Bergen, Breck tried his first practical demonstration of the methods of oceanography by demonstrating, to those guests who were interested, the age-old method of measuring the speed of ocean currents. It had been a calm day and Captain Athanasios had given permission for the ship to be stopped and a large, orange sea-anchor somewhat like a large partly deflated balloon had been thrown off the stern to keep the ship stationary. Breck had explained to his audience about the old methods of throwing a "drifter" or float overboard and then measuring the rate of its progress past the ship in order to measure the local surface current. It was not unlike the similar method he explained in measuring a ship's speed by throwing a "log" overboard and measuring the number of knots slipping past on a line loosely held by a crewman. What the guests did not know was that the Mystery was being held in a very exact position by thrusters – ducted propellers in the sides of the ship at the bow and at the stern – which kept her very stationary at a precise GPS position already planned by Breck and his assistants in advance. Whilst he went through the elementary demonstration of how to measure an ocean surface current at the stern of the ship Breck knew that his two assistants from the crew, Fionn and Rónán in the laboratory below deck had already been collecting the data. They had been doing this since leaving Bergen and whilst the ship had been underway using the most modern acoustic Doppler profiler device to accurately measure the ocean current near the sea floor. The Acoustic Doppler Current Profiler, attached to the bottom of the ship sent an acoustic signal down towards the sea floor so that the sound waves bounce off particles in the water very similar to the less sophisticated depth recorder. The speed and direction of the current is found by knowing the frequency of the return signal, the distance it travelled, and the time it took for the signal to return to the ship's sensors.

Breck's own simple demonstrations had gone down well and was enjoyed by the majority of guests who had assembled at the scheduled time on the afterdeck of the ship. Even the two Russians had shown considerable interest and appeared to have become more sociable with Breck and the other guests. Breck had explained the purpose and basic history of measuring surface currents, referring to the Norway Current in which they were now travelling. This current, he explained ran northwards along the Norwegian coastline and was itself a branch of the North Atlantic Current, sometimes considered a continuation of the warm Gulf Stream which originated in the tropical waters of the Caribbean.

To Breck's mind the day had gone well. He had successfully given his first practical demonstration of one of the simpler methods of measuring surface currents. Fionn and Rónán in the laboratory below deck had also been gathering a considerable amount of data from the bottom counter current and Fraser had given a very entertaining lecture in his delightful Scottish brogue on the variety of marine mammals which may be encountered on their voyage. Breck had also received generous praise during the usual socialising in the ship's lounge later in the evening, so he went to his cabin content and rather sleepy after enjoying a few "wee drams" with his colleague, Fraser.

Breck suddenly woke up. Something was not quite right. He switched on his bedside lamp and looked at his watch. It was now dark outside, the sun not setting until well after ten in the evening in these high latitudes. Two in the morning. What was it that made him restless? His many sea voyages had made him accustomed to the usual shipboard sounds: the rustle of the water going along the sides of the ship; the occasional bump as the ship met with a wave from another direction; or the sound of a high wind in the superstructure of the ship. Something now was different, and he could not place what it was. He climbed out of his warm bunk and put his dressing gown over his pyjamas, left his cabin and went through the nearby double doorway and out onto the upper deck. It was a cold night, but the sky was clear, and the multitude of stars above were reflected in a glassy sea which rolled gently like undulating black hills set with many small sparkling gemstones. Only the phosphorescent bow wave of the ship disturbed this beauty. He walked down towards the stern of the ship, but stopped suddenly when he saw the darkened figure of a man at the railing silhouetted against the sea. He had a flashlight and was sending intermittent flashes to the darkness beyond the ship's stern. Morse Code!

Before he could come up to the man, he felt a small object dig deeply into his back and a quiet but threatening, strongly accented voice close to his ear said: "Not a sound, Dr. Breck. This is only a small pistol, but it will still kill you where it is. Move forward to my friend at the railing."

The man at the railing suddenly stopped his signalling and turned. The voice behind continued quietly "Dr Breck is joining us for a little while,

Yuri. I think that he needs to study his oceans more closely, yes?" Breck was pushed forward against the railing and was grabbed by the other man.

Suddenly there was another voice, a louder voice with more authority to it: "I think not, gentleman." The light from an opened cabin door showed that the captain, Calypso and several of the crew had come onto the deck and now confronted the two Russians and their captive. To Breck, everything then seemed to happen very quickly. A crewman rushed forward and pulled Breck back towards the captain. There was a loud 'crack' as the pistol in the Russian's hand fired. Captain Athanasios staggered back, more in shock than from the impact of the small calibre bullet from the pistol and a small red spot, which quickly spread wider, appeared high on the left breast of his white uniform shirt. *A fatal wound* Breck thought.

Calypso thrust out her arm towards the two Russians and the crewmen leapt forward, knocking the weapon from the Russian and lifting them both up upon their shoulders with little effort, rushed to the railing and threw the two men overboard into the sea.

The crewmen quickly returned and hurriedly helped their captain back through the open cabin doorway. Calypso pulled a lifebuoy from its mounting on the railing and threw it overboard after the two Russians. She then took up the flashlight which the Russian had dropped, pointed it out towards the stern of the ship and flashed a Morse Code signal:

-.-- --- ..- .-. / -- . -. / .. -. / ... . .- / - --- / .-- . .. -.. -.- / ..- .--. [10]

Coming up close the Breck, she handed him the flashlight, looked directly into his eyes and said:" Not a word to anyone, Dr. Breck. Is that understood?"

The look in her eyes told Breck that she was now the First Officer and, with her captain incapacitated, she was now in command and that

---

[10] Your men in sea to pick up

there was no room for mercy. She followed the last of the crewmen through the doorway and closed the door, leaving Breck once more alone on the darkened deck. It took him a little while to come to the realisation of what he had just witnessed. He turned and looked out to sea and saw in the distance a thin beam of light scanning the ocean's surface.

Curious, he picked up the discarded flashlight, switched it on and began his own search of the metal deck. He saw a small, white object further along the deck and walked over and picked it up. It was the small pistol which Kavalenko, the Russian had pushed into his back and then later used it to shoot the captain. He quickly put it into the side pocket of his dressing gown and returned to his cabin.

Back in his cabin, he switched on the main overhead light and looked carefully at the small pistol which had obviously went through the ship's security scanners undetected. It was about the size of his hand and consisted of a small, metal barrel and a main body made from a hard, white plastic which had been made using a 3D printer. Breck found that he could unscrew the metal barrel and open the body by pressing securing tabs around its edges. Inside he found that the firing mechanism also consisted of a metal firing pin, spring, and also a small hard plastic trigger. It was a single shot weapon but set into the hand grip of the weapon was a compartment which now contained four spare .22 calibre cartridges. The metal barrel, when unscrewed from the body, could easily be disguised as an ornate fountain pen. The device had obviously been smuggled in through the security scanner, being several innocuous and undetectable plastic parts, whilst the metal barrel containing all five cartridges, would have been detected as a simple "fountain pen". 'Ingenious' thought Breck. Without thinking any further, he reassembled the weapon and put it into his bedside table drawer. He was confused as to what he should do. He thought that perhaps that he should he go and find Calypso and the captain but then thought that she and the crew would be more occupied with Captain Athanasios, if he still lived, and would probably resent any interference from an outsider. Still with considerable apprehension on his mind, he switched off the main light and climbed into bed. Sleep was impossible and he stared at the ceiling of the cabin, his thoughts wandering over the events of his voyage so far on the *Mystery of the Seas.*

Chapter 10: The Maelstrom

Sleep did not come easily for Breck that night. The events of the last few days were on his mind, and he was greatly apprehensive about the future. The sudden tsunami at his father-in-law's place in Florida; the convenient invitation to this voyage; the death of Ziminov and the attack of the two thugs in Bergen; and now the shooting of Captain Athanasios all seemed almost too much for his comprehension. When sleep did come, it came suddenly and deeply and Breck woke with a start, the morning sunlight coming through his single porthole. Sunrise had come many hours before in these northern latitudes, but he looked at his wristwatch and found that it was much later his usual time for breakfast. He quickly got out of bed, put on his clothes and hurried out of his cabin and headed for the dining room.

When he arrived, he found that most of the guests had finished but he was totally shocked and confused when he saw Captain Athanasios and

Calypso sitting at the end table laughing over some trivial matter. The captain was in his usual white uniform of jacket and slacks and showed no sign of his distress a few hours earlier. As Breck came up to the table and took a seat opposite Calypso, she looked up, smiled and put a finger across her lips. Captain Athanasios turned to Breck and said jovially: "Good morning, Dr. Breck. I hoped that you slept well. Perhaps after breakfast you and Calypso might come to my cabin for a quick chat?" With that, he stood up, wished them both a "good morning" and left; walking through the dining room and stopping occasionally to talk to some of the guests. Breck looked at Calypso who simply raised an eyebrow and went back to finishing her breakfast.

Somehow, Breck managed to finish his light breakfast of hot buttered toast and black coffee; his mind was now in a new state of excitement; he had felt sure that a gunshot to the upper chest would have finished most men, even with such a small calibre pistol as the one which he now kept in his bedside drawer. Calypso had finished but went to the servery for another cup of coffee and returned to the table. She said nothing, leaving Breck to his confused thoughts. Again, he wondered what he had gotten into. He was living in two worlds, one of affluence and social prestige as a guest Presenter aboard a luxury cruise ship and the other of death, secrecy and confusion.

He finished his breakfast, putting his empty cup back on its white saucer with the blue trident monogram. Calypso stood up and put her hand on his arm saying, "It's time to visit the captain. Are you ready?"

Breck simply smiled weakly and stood up slowly from the table. He was both apprehensive and curious as to what Captain Athanasios had to say about the previous night's events. He followed the tall First Officer out of the dining room and into the elevator which led to the deck on which the captain's cabin was located. She knocked on the door and they both entered. Captain Athanasios stood up behind his desk and motioned that they should both sit down in the two armchairs which had been placed in front of the desk. He sat down and raising both hands slightly in a friendly gesture of depreciation.

"Well now, Dr. Breck. It was a most exciting evening was it not?" the captain said with a smile. "As you can see, I am perfectly well. That little Russian 'peashooter' hardly raised a scratch and besides, our Medical Orderly, Damon has remarkable powers in healing."

Breck sat quietly; he was not sure as to what he should say under the circumstances. The captain continued, giving a quick glance at Calypso; "And please don't be concerned over my First Officer's instinctive reaction to remove the source of the threat to her captain. You will be pleased to know that all of us on board the *Mystery* value human life and that she ensured that our two spies were picked up by their own people. This was confirmed by my lookouts who used night vision binocular and saw that the ship, which had been following us, did launch a boat and was able to locate the two men clinging to the ring which Calypso had thrown after them. How do you feel now, Dr. Breck?"

Breck moved uneasily in his chair and quickly glanced at Calypso who was sitting quietly next to him, her hands lightly clasped together on her lap and her face looking towards the captain. "Rather uncertain I must say, Captain Athanasios. I have had a few exciting moments in my work as an oceanographer, especially in the Antarctic, but this voyage so far has given me a few concerns."

The big captain stood up and walked around to the end of the desk so that he stood next to Breck. He leaned down and put his hand on Breck's shoulder. "Do not be concerned, Terence. Our voyage should be more mundane and normal from now on. Our 'Ukrainian' guests have left us, although their ship still follows us just out of sight. A search of their belongings this morning showed that their names were Morozov and Kozlov and that they were ex-military rather than SFB and are now employed by Sea Wolf Shipping Lines, a Russian public company controlled entirely by one Dimitri Alexandrovich Volkov. At least we know our enemy and why he would want to suppress Dr. Ziminov's theory of new Artic freezing as it would put an end to his shipping prospects. Volkov is a very powerful oligarch with a direct line to the Kremlin. My organisation has known about him for some time as he seems to have a distinct disregard for the pollution which his ships cause to our seas, especially in the Baltic and now in the Arctic. He is a dangerous man and has no problems in using his security forces to silence his opposition. But, getting back to reality, our other guests will be told that our former 'Ukrainian' passengers have been urgently called back to their own country and have been taken ashore by a local pilot vessel – perhaps not a bending of the truth too much" he laughed. "But do not worry, your work is very important to us, and I am sure that the results will be vital for the future of the world's understanding of climate change. Please put last night's excitement behind you; we have much more important work to do, and we will soon rid ourselves of our persistent friends. Do you have any more questions?"

Breck felt reassured by the use of his first name and stood up and faced the captain who seemed to be larger and more protective at such close proximity. "No sir." Was all he could meekly say.

"Good!" said the captain who walked smartly over and opened his cabin door. "The sun is not yet over the yardarm, as they used to say in the old navy, and I am sure that my good First Officer would like to take you down to our comfortable lounge and buy you an early morning scotch - Johnnie Walker Black Label 12 years old is your recently required drink of preference, is it not?" he said with a smile.

Breck gave the big captain a weak smile and followed Calypso out of the cabin. Further along the companion way leading to the elevator,

Calypso turned slightly as she walked, smiled and said "It seems that our former Russian guests were being told to get the documents which we obtained at the hotel in Bergen. I think that this is not the end of the matter, but do not worry as Poseidon always takes great care of his crew and has great capabilities. You have nothing more to worry about, I am sure. So! Let us have a quick drink to calm our nerves and get back to work." She walked on.

Breck accepted the glass of scotch in the almost empty lounge and sat quietly. Calypso gave a little laugh at his silence and put her hand on his arm. "Do not worry, Terence. All is well now. We will lose our tormentors and get on with our duties because they are so very important".

Breck smiled and took the last, quick gulp of his whiskey. "Yes, of course. You're right, Calypso. I have much to do and much to forget also." He stood up and gave her a mock salute before heading towards the elevator and its entrance to the ship's secret facilities well below. His dream of a happy working holiday aboard a luxury ecotourism ship had become a nightmare in which he had been forced to exist in two realities: one as a genial guest speaker showing the passengers the simpler aspects of his favourite occupation; and the other as a clandestine investigator involved in death and subterfuge and secret research into what could be a global catastrophe.

As he descended in the elevator, passing the public decks of the ship, he also found it difficult to believe that these covert activities aboard the ship and ashore had been successfully kept from the other passengers. The officers and crew of the *Mystery* were devoted to their captain and went about their duties as any crew aboard a cruise ship would. Even the occasional short stop at sea to allow him to demonstrate some simple 'On Station' technique of oceanography to the passengers had another function – allowing his two assistants below in the oceanographic laboratory to do more meaningful depth soundings and bottom current measurements. Meanwhile, he had also played his part as a guest on board this deceptive ship. It was easy to socialise with the other passengers, become involved in the nightly lectures and dinners and to show everyone that he was just a simple oceanographer taking some well-earned leave to pass on his knowledge on an interesting ecotour along the Norwegian coast and into the Arctic Ocean.

The captain had been right and life aboard the *Mystery* had returned to some form of normality for Breck but on their tenth day out from Edinburgh, there was a knock on his cabin door very early in the morning. A glimmer of dawn was beginning to filter through his cabin porthole, and he looked at his wristwatch. Five ten. Not a time for any of the guests to be awake and well before the end of the Morning Watch at 8 am for the crew. He quickly put on his robe which he had slung on a hook near his bed and went to the door. He was surprised to find Calypso the First Officer standing there in her usual immaculate white uniform.

She held a finger up to her lips and said quietly "Come! There is something which you might like to see. Quickly now!" She quickly turned and walked off smartly down the companionway. Breck tightened the cords to his robe and followed her, barefoot and cold through the silent ship. Eventually she led him up to the bridge where it seemed dim and deserted except for a lone crewman at the helm. Looking around, he saw the captain standing alone far out at the end of the Port Bridge Wing. Unusually, he was facing astern and had his arms folded across his broad chest. His head was raised and his eyes were closed but his lips moved slightly as though he was intoning a silent prayer.

"No! Leave him!" Calypso said quietly and turned and walked out to the other exposed Bridge Wing. Breck had regretted that he did not take the time to dress more appropriately as the morning air was very cold. He joined the First Officer at the far end of the wing. She had taken a large pair of binoculars out of a small locker there and handed them to Breck. "Here! Look astern slightly to the left of that distant headland – about five degrees."

Breck took the glasses, rested his elbows on the edge of the wing's surrounds and focussed his view in roughly the position that Calypso had given. There was a small ship, bow on and coming in their direction.

"Our Russian friends." Calypso said with some disgust. They have been following us since we left Bergen as you know. Usually, they have kept out of sight but these waters are treacherous and so they show less concern about being seen."

"What are they up to do you think?" Breck said putting down the binoculars.

"Nothing much at the moment but soon they will be in real trouble and have to show some real seamanship!" she said with a wry smile.

Breck gave her a quizzical look, so she continued, extending her arm out towards the distant headland. "That strip of land is the end of Lofoten Point and we are about to turn into the board strip of sea between the Lofoten Islands and the Norwegian mainland. The Russians unfortunately are in the wrong bit sea and at the wrong time."

Breck had heard this name before, and he searched his mind for the connection. It suddenly came to him in all of its horror. "The Maelstrom!" he said with awe. "The giant whirlpool which sinks ships is off the Lofoten Islands!"

Calypso calmly took up the binoculars and looked out towards the Russian ship. "You have been reading too much of Edger Allen Poe," she said with a small laugh. "Probably not as dramatic as a sinking but soon the Russians are going to have some fancy navigation to perform. Captain Athanasios has been very accurate in getting the *Mystery* to our current position, knowing that the Russians would come closer. The tide is about to turn, and the wind is coming from the right direction and strengthening. Also, the Russians have followed us exactly and so they are now over a part of the sea floor which will enhance the coming Moskstraumen – what the Norwegians call this particularly violent maelstrom whirlpool. Look!" She handed the glasses over to Breck.

Even at this distance, Breck watched with grim fascination as the sea around the small Russian ship began to toss and boil, turning it now side on to his view.

"Do not be too concerned for them" Calypso said without pity. "I am sure that they have the necessary skills to get out of trouble. If they are smart, they will follow the direction of the water flow until they can gradually escape. Which way are they headed?" She asked.

Breck gripped the binoculars more tightly mostly due to the stress of the situation more than to simply get a steady focus, "They have turned north towards the land!"

"Good! I know these waters very well. The tidal current will carry them up to the island of Rødøya where there is a safe inlet, but they will probably run aground there as it is very shallow. I think that we can say that we have finally shaken off our Russian wolfhounds. And tomorrow our guests can have a safe excursion in the lovely seaport of Reine." She turned and left the wing, leaving Breck trying to come to terms with this new mystery of the seas.

Chapter 11: A New Threat

Dimitri Alexandrovich Volkov, owner of Sea Wolf Shipping, sat in his plush office on the fifth floor of the Azimut Hotel on Prospekt Lenina at the Five Corners Square – the main square of Murmansk. He turned his chair around and gazed out of the big picture window across the city and dockyards. Most of the buildings were blocky, Soviet-era constructions of less than five stories and he felt smug in his office in the tallest building in the city. The first five floors of this luxury hotel had been devoted to office space and his was the best in the building and faced towards the northwest, overlooking the docks of Kol'skiy Zaliv[11] where his latest ship, the nuclear-powered icebreaker the *Vasili Brusilov* was docked.

He had just finished the report sent to him via the company's secure network, that the threat posed by the climatologist Dr. Ziminov had been reduced somewhat by the 'unfortunate accident' which the

---

[11] Kola Bay

good doctor had had in falling out of his hotel window whilst trying to see fireworks. He smiled when he read that Ustrashkin's two agents Komarov and Gusev had been able to assist Ziminov in his endeavours. Unfortunately, they had not been able to obtain Ziminov's diary and papers which he had left at the hotel's front desk with very specific instructions to give the documents only to a Dr. Terence Breck, an oceanographer who also was aboard the ship, the *Mystery of the Seas*, and a highly qualified oceanographer who worked for the American government. The two agents had apologised about this matter, stating that they had attempted to take the documents from this American but were intercepted and beaten up by the crew of the ship. This was only a minor setback as Volkov knew that there were still two agents on board the *Mystery* who would be instructed to steal the documents from the American's cabin. Moreover, his small supply vessel, the *Vladimir Alafuzov*, had been ordered to follow the *Mystery* out of Bergen to assist the two agents and to instruct them to obtain Ziminov's documents or sink the ship.

There was a firm knock at the door and Volkov's friend, Head of Security and general 'fixer', Sergey Ustrashkin, entered without waiting for the customary invitation to do so. He seemed agitated.

"Bad news, Dimitri Alexandrovich!" he said, waving a piece of paper in his hand and walking quickly up to Volkov's desk. Captain Orlikov of the *Alafuzov* has just sent us another signal. Not good, I am afraid."

Volkov leaned forward onto his desk and folded his arms. "Well! Get on with it, man. What is this 'bad news' from our Captain Orlikov?"

"When he tried to contact our two agents aboard the American's ship, they were discovered and thrown overboard by the ship's crew." As well, the *Alafuzov* has run aground and cannot continue their surveillance of the *Mystery*." Ustrashkin saw the anger well up in Volkov's face who now stood up suddenly, both his fists clenched.

"But there is some good news coming from it!" Ustrashkin continued, "they shot and killed the captain of the *Mystery* before they were captured.

"Perhaps." Volkov said, sitting down and looking at the report on his desk. "But they have Ziminov's papers and I believe that there is an American oceanographer aboard who may be able to make some sense of it."

"If he can speak Russian!" Ustrashkin laughed.

"Considering the bungled attack by our two idiots in Bergen, I am sure that the American will soon find a translator. Remember, should anyone take Ziminov's theories seriously and publicly state that there is even the mere chance that the ice will again return to the Arctic Ocean, then our shares will fall through the floor. Moreover, we will no longer get any more preferential treatment from our 'Grandpa in his Bunker' at the Kremlin and we may even disappear like some of our colleagues. Do you want that, Sergey?"

Ustrashkin stood silently for a moment, considering the fragile nature of Volkov's company and possibly their lives should their plans to use the open Arctic Ocean as a regular shipping route between Russia and North Korea and other friendly nations failed. The Western blockade of Russia had been tightening over the last year and the Kremlin was becoming impatient. "So, what do we do now, Dimitri Alexandrovich?"

Volkov picked up his cell phone and made a call to the dockyard where his icebreaker, the *Vasili Brusilov* was moored. He spoke for a while and then put his phone back on the desk and looked at his friend and Head of Security. "Get some of your lads together, Sergey. Full kit and plenty of explosive charges. Captain Khovrin's ship is able to sail, and I will have the good captain report for duty first thing in the morning. It looks like we will have to go after Ziminov's documents ourselves, even if we have to sink their silly little ship. Remember the old saying Sergey, "one's own shirt is closer to the body.[12]"

Captain Ivan Khovrin scowled as he walked up the gangway to his ship, the Nuclear-powered icebreaker *Vasili Brusilov*. It was still only six in the morning, but the sun was already up as a dim glow through the cloud cover low on the horizon of the low hills which surrounds Murmansk. A cold wind was blowing across the grey waters of the Kolský Záliv, that narrow inlet which snakes southwest from the Barents Sea through the low hills upon which the city was built, heralding a cold winter yet to come.

He scowled because it was early in the morning and his extended leave from sea duty had already begun and because he saw that coarse thug Ustrashkin and a small team of his security team,

---

[12] "Своя́ руба́шка бли́же к те́лу" This saying emphasizes the importance of taking care of oneself and one's own interests first.

armed and in full combat gear coming along the dock. *"What a dreadful start to the day!"* he thought as he continued up the gangway and onto the deck of his ship. *"His ship. Ha!"* he thought as he climbed the stairs up to his cabin on the Bridge Deck. Since that upstart Dimitri Alexandrovich Volkov had taken over from his father, the company had become a mere private plaything of the Kremlin with violence being the usual form of corporate negotiation.

Captain Khovrin opened his cabin door and was surprised to find Volkov sitting in his captain's chair. "Ah! Good morning, Captain Khovrin. I was wondering when you would eventually get here." Volkov said sarcastically. Standing up, he walked around to the large wall map which was on the opposite wall. "We have a rather unfortunate situation to deal with and so I need you and the *Brusilov* for some extra duty."

"This is most outrageous, Mr Volkov!" the captain protested. It was rare that he used the honorific for the other man as the captain thought very little of his superior and usually gave him no title to his face. With trusted crewmen, however he called him 'the little Czar,' behind his back and knew that very few people in the company liked their employer. His protests that both he and the crew of the *Vasili Brusilov* were just about to start a week's leave after the ship's recent refit, fell on deaf ears as Volkov picked up a pair of dividers and measured a distance across the top of the map across the Barents Sea and up to the Norwegian island of Svalbard. "About eight hundred nautical miles, I think. About three days sailing. But I think that we will find our quarry well before then if they venture into the Barents Sea proper. What do you think, captain?"

Khovrin turned and looked at the map and shrugged his shoulders. "Probably" he said with little enthusiasm. He knew that despite his feelings towards the man, the head of Volkov Shipping had done his apprenticeship under his father and knew the basics of seamanship. "But what is the reason for this unprecedented sailing?"

Volkov opened the cabin door to leave, but quickly turned and said sharply, "The details are not for you to be concerned about, captain, except that if we fail, you may be out of a job. Just take your ship out into the Barents and head northwest. Have your best man at the radar and its limits extended as far as it allows. We are after a small ship – under about 3000 tonnes and it will probably be the only one of its size in these waters at this time of year.

Chapter 12: Into the Arctic Ocean

The journey of the *Mystery of the Seas* continued northward along the Norwegian coast. Breck had relaxed somewhat since his last encounter with the two Russians and had now immersed himself into his secondary life as a mere guest speaker in oceanography. So far, Poseidon and his crew had managed to keep their research activities, their contacts with the Russians and their ship from the other guests. Even the other guest speaker, the marine biologist, Hamish Fraser had no idea of the dramatic events which had taken place since their arrival in Bergen, although he had remarked about Breck's absences from the daily social events on board. Breck had excused himself as being a rather shy type who needed time to prepare his notes and lectures more thoroughly than simply being a practicing oceanographer. He did, however, enjoy the Scotsman's company and now spent more time and a few 'wee drams' in the evening discussing the man's work and the customs of the country of his own ancestors.

Breck was now able to put in more time in studying Ziminov's notes, ably translated by Armenus, the Russian-speaking Chief Steward, and also the data which Fionn and Rónán had obtained from their clandestine sampling of the bottom currents along the coast. Both studies were disturbing.

Ziminov's data suggested that some parts of northern Siberia have been witnessing increased snow cover over the past decades, even as rising air temperatures was already melting glaciers and polar ice caps in an interesting paradox. A search of the more reliable scientific sites of the Internet also showed similar findings from some other researchers who had also noted that the melting of the northern ice cap and the increased evaporation of the sea in the Arctic Ocean were probably related to this increase in snowfall[13].

---

[13] Real research and data from:
https://www.nature.com/articles/s41612-022-00310-1
https://www.amap.no/documents/doc/long-term-variation-in-snow-depth-and-duration-in-northern-eurasia/956 and
https://iopscience.iop.org/article/10.1088/1748-9326/4/4/045026

Moreover, Breck's own studies aboard the *Mystery* did show a weakening of the counter current flowing south below the Norwegian Current at the surface. If this continues Breck knew, there would be a reduction or even a collapse of the warm Norwegian Current and Gulf Stream which brought warm water from the tropics to the coasts of western Europe and Great Britain.[14] Without the influence of these warm currents, the temperature of these northern regions would suddenly drop to be more like the coasts of the Canadian North Atlantic. Europe would then experience very low temperatures, and with Ziminov's findings of increased snowfall in Siberia, his hypothesis of an imminent new Ice Age could well become fact.

---

[14] Real research and data from:
https://www.scientificamerican.com/article/if-the-atlantic-ocean-loses-circulation-what-happens-next/
https://www.nature.com/articles/s41467-023-39810-w and
https://insideclimatenews.org/news/09022024/climate-impacts-from-collapse-of-atlantic-meridional-overturning-current-could-be-worse-than-expected/

Breck had gone over the data and the reviews of the published literature and had rechecked Ziminov's data and hypothesis several times. The last few days had brought home the necessity and urgency of the voyage of the *Mystery of the Seas* and of Poseidon's own apprehensions.

He was going over his own data from the Norwegian counter current yet again, looking for some fault which might prove Ziminov's hypothesis incorrect when there was a knock at his cabin door. Opening it revealed the tall First Officer. Calypso had a worried look on her face and said quickly. "Come, Terence. There is something you should see on the bridge." She turned and walked quickly off down the companionway. Breck followed her as fast as he could until they arrived at the bridge. Captain Athanasios was over at the starboard side of the bridge looking out to sea through his binoculars. He sensed their approach and turned, saying grimly "Well, Calypso! I think that we have more trouble from our Russian friends." He handed the glasses over to his First Officer. She took then and looked in the direction which the captain indicated. "Another ship. Large. Perhaps an icebreaker."

"Exactly!" Poseidon replied. It is the *Vasili Brusilov*. Nuclear powered and coming out of Murmansk, I think. She is owned by our old friend Dimitri Alexandrovich Volkov owner of Sea Wolf Shipping. She has been coming towards us for some time and I do not feel that her intensions are friendly. We have just rounded the Knivskjelodden Peninsular, the most northern point of Europe and as you can see, we are amongst some pack ice close to shore. I think that this is about as far east as I wish to go, so we will turn north and continue on to our destination at Longyearbyen in Svalbard." He turned towards Breck: "Have you enough data, Dr. Breck?"

Breck took his gaze away from the distant ship and reply "Yes, thank you, sir. The results of our work and the data from Dr. Ziminov's notes are all that I need. Their implications are quite alarming."

"Thank you, Dr. Breck. I think that our present position may be more alarming still and I must ask you to go below as we now have some manoeuvring to do on the bridge to get clear of the pack ice and that Russian ship. She is nuclear powered, very much larger than us and much faster. We may be lucky and lose her in a sea fog which is common in these waters at this time of year." He said as he walked out onto the open starboard bridge wing.

Breck gave an apprehensive look at Calypso who had turned and was now looking at their current bearing on the gyrocompass. As he was leaving the bridge, he heard her give the order to the helmsman "Steer bearing 300 degrees." They were turning north and Breck hoped that they could escape the Russian icebreaker.

By the time Breck had reached the main deck, he found Fraser and some of the other passengers looking astern at the distant ship, now becoming larger and gleaming white amidst the many white and blue ice flows covering the sea in the afternoon sun.

"Ah see that we hae some company th'day. A welcoe chaynge fae days o' open sea" he said in his deep Scottish brogue.

Breck came up to his friend at the railing and looked out at the approaching ship.

"Yes, indeed Hamish. But I feel that she may be unfriendly." He said without giving away the apprehension of those on the bridge.

"Aye. That micht be sae. Th' Ruskies dinnae lik' fowk comin' intae thair backyard," the Scot said turning to go back inside but stopped and pointed out to sea past the starboard bow, "Ken thare,

Terence! Thir's a muckle fog bank comin' up. We micht lose thaim in that!"

The cloudy white wall of the fog bank soon enveloped the *Mystery* in a clammy, translucent world where the sun became but a distant red glow and where visibility was no more than about twenty metres. Breck was surprised when he felt the ship slow down and eventually stop. The bow wave which usually ran along the side of the ship had ceased to flow.

"Whit noo, Terence? ur we aff tae anchor 'ere in th' open sea?" Fraser said, turning to Breck.

"I think not Hamish, the Barents Sea here is very deep, I believe, but there doesn't appear to be much drift, and I think that the captain is hoping that the other ship will pass us by. Also, whilst the sea in front seems to be generally free of ice, we only have to hit one of the larger growlers in this fog to do ourselves a mischief.

"But whit aboot radar? Wilnae he'll see us wi' that?" Fraser said with a little anxiety.

"Oh, he'll see us all right Hamish, but in this fog, he is also likely to stop and wait until the fog clears. We are only an 'ice-strengthened' ship, so we need to stop in this pack ice. Their ship has a bow which will ride up and crush the ice, but their propellers are subject to damage with any thick lumps of ice and if the flows are thick enough, they may also damage the weaker sides of the ship. No. I think that they will stop also. I'll go up to the bridge and see what is happening."

Breck went back to the bridge and was met by Calypso, who was not happy to see him, but understood Breck's anxiety about the motion of the Russian ship. She confirmed that the *Vasili Brusilov* did continue for a while, coming up close to the *Mystery* so that she was only a little more than a kilometre off their stern. Both ships were now drifting amongst the ice flows of the Barents Sea close inshore to the northern coast of Norway.

Breck returned to where Fraser was waiting at the stern railing, and told him that the Russians had also stopped and were somewhere off their stern in the murk of the dense sea fog. They chatted for a while about what the Russians wanted; Breck trying not to give his friend the real purpose of the Russian's presence. It was deathly silent in the fog but suddenly there was a splash of water below and they both looked over the railing and into the grey-green glassy sea below. A small black head resurfaced, its whiskered face looking up at the two men. A second dark grey head surface momentarily and then both were gone beneath the dark water.

"Ken thare!" Fraser cried pointing down at the swirling pool where the small head had been. "If I'm no mistaken that wis a wee grey seal!"

Breck looked over the side but could not see anything. He looked up at his friend who shrugged his shoulders and smiled. "Ah hae heard that grey seals wur found in these waters 'n' I'm thinkin' that un wis a micht curious as tae how come we ur 'ere." He looked back over the side and then continued, "Mah auld granny back hame at Stromness in th' Orkneys used tae tell me aboot th' seals changing intae folk – 'Selkies' she cried thaim - a wee bit mair interesting than mah studies o' th' grey seal a' tha' university." He laughed.

Chapter 13: Death in an Icy Sea

On board the *Vasili Brusilov,* Volkov was not a happy man. He had watched with anticipation as the blip on the ship's powerful radar finally resolved into a small image of the real vessel in his binoculars. It was a small ship with a blue hull and white superstructure; an insignificant ship in Volkov's perspective.

"Bring your ship up against that one." Volkov said to Captain Khovrin who stood impassively further over on the bridge and who was also looking at the *Mystery* through his binoculars.

"That is not going to be easy." The captain replied without lowering his glasses. "The pack ice has increased and there is a fog bank coming in from the north which will soon engulf them."

"Get as close as you can, and I will send Ustrashkin and his team over in a Zodiac to board them." Volkov replied with some irritation in his voice.

"I would not advise that, sir." Captain Khovrin said, pointing out towards the smaller ship. "The fog bank is coming quickly and is already engulfing that ship. Your men will get lost trying to get to her."

Volkov did not reply but stormed out onto the bridge wing of the icebreaker and then returned, a look of malevolence on his face. "You have radar. Full ahead and ram that ship!"

Captain Khovrin looked up, startled at his superior's orders. "That is impossible Mr. Volkov! We are an icebreaker built for commercial and peaceful operations. I cannot do such a horrid thing against another ship in these waters. That would be an act of piracy!"

Volkov was now in a rage "Get closer then so we can board her. She is a threat to the company and your future, captain. Do it!"

The captain waved his hand across towards the bridge windows which were now becoming obscured as the sea fog began to cover the ship.

"That will be impossible now, Mr. Volkov. Even with radar, we will not see the larger ice flows which will hit us, and which may cause us some damage. No! I am sorry. We will have to stop until the fog clears and then we will call upon that ship to stop. If you wish, I could make up an excuse about being in Russian waters, but I think that their captain would not fall for that. I am sorry but we must stop."

Volkov pushed the captain aside and took over command of his company's ship. They continued through the fog; deep thuds were heard as the hull encountered some of the bigger ice flows and the ship's speed was noticeably reduced. Eventually Volkov listened to the captain's protests and left the bridge. Captain Khovrin then ordered the *Vasili Brusilov* to stop.

The next morning, the sun had only just risen in a clear eastern sky when Volkov returned to the bridge. Captain Khovrin was still asleep in his day cabin nearby and Volkov picked up his binoculars and looked out towards the other ship which was now clearly seen and seemed to lay only about a kilometre off their Port bow.

"Good!" he said to himself and then turned to the officer on watch. "Get this ship going and head towards them." He said pointing towards the *Mystery* which seemed to lay helpless in the calm water beyond.

The officer rang the telegraph to SLOW AHEAD for the engine room below and directed the helmsman who was fallen asleep on a nearby chair to once more take the wheel. There was a shuddering of the engine below and a vibration which the officer though was unusual. The ship did not move.

Captain Khovrin was now awake. A good seaman, he was very much attuned to every wave which struck the hull and any other noise other than what should be occurring during normal operations. "What's going on?" he demanded, doing up the last button of his sea jacket. "What's that noise?"

The officer in charge looked at his captain with a blank expression on his face. "I am not sure, captain. It sounds like we have fouled our propellers, and we are not making any headway."

"Stop the engines!" the captain shouted at the Quartermaster at the controls. "What are you doing to my ship, Volkov?" he said in some anger to his superior and rushed out to look astern from the end of the bridge wing. The ship had stopped and there was only a confused swell coming out from the water at its stern. "We are fouled!" yelled

the captain back to his staff on the bridge. "Get a diver over and see what it is immediately!" he said returning to the warmth of the bridge.

"How did this happen?" Volkov said angrily. "What can foul our propellers out here?"

"Who knows? Kelp, probably. There are some massive undersea forests of the stuff inshore, even in these cold waters."

It took a while for the ship's diver to dress in his warm Dry Suit and scuba gear and be lowered over the stern railing to the icy water below. He was only submerged for a few minutes. He looked up at the captain and Volkov who had come aft, and he raised his face mask and yelled up to the men. "Rope! Both props are entangled in thick mooring rope. It will take hours to cut it all free."

"Rope!" yelled Volkov "How could we run afoul of rope here in the open Barents Sea?" he said angrily turning towards the captain who looked perplexed.

"Jetsam, I can only suppose" he said lamely, shrugging his shoulders and looking down at the diver who was now being hauled back on board. "Thrown overboard or lost from some other vessel. Who knows?"

Volkov quickly walked over to the bridge telephone and aggressively punched in a number. There was a delay much longer than he anticipated, so when there was an unintelligible "grunt!" at the other end, he yelled into the speaker "Ustrashkin! Get your team ready, now! Full boarding equipment and assemble at the Zodiac crane. I want you to cross over and board this small ship. Eliminate everyone on board and set you charges. I want no evidence of its existence. Clear?"

There was another small delay and then a startled response, "All of them, Dimitri Alexandrovich?"

" Yes! Now do it!"

Captain Khovrin strode over to Volkov and put a large, calloused had upon his shoulder to turn him around. "You can't do that! That would be murder, and we are not murders here on my ship." Volkov shrugged it off and faced his captain. "I can easily replace you now!" he said, drawing a small automatic pistol from his jacket pocket. "Or you can go back to your cabin and think about your pension and family and keep quiet. There will be no record of this event now or always, captain. I have many ears back home and my reach goes out well past you and your family. Think of that! Now get out of the bridge and let me get on with saving the company."

Captain Khovrin looked at his Watch Officer and the Quartermaster who stood rigidly at their positions on the bridge, pretending to hear nothing of the tense conversation between the ship's captain and its owner. Khovrin shook his head and left the bridge.

Volkov walked out to the end of the bridge wing and watched Ustrashkin and his team of five well-armed men being lowered in one of the ships sleek, black Zodiac inflatables; Ustrashkin himself at the controls. He turned and looked over the Port bow at the other ship. "*Why had she not moved?*" he thought. There is enough light to get underway and the danger would have been obvious. He looked back at the Zodiac, now motoring its way, skirting the many irregularly-shaped flat ice packs which stood between them and the smaller ship and trying to find the few sections of open water. Volkov took the binoculars from their bridge wing case and looked out towards the Zodiac which was approaching the other ship. "The fools!" Volkov said to himself. "Why are they waiting?"

He soon found the answer to his question in a most horrific way.

He watched with some alarm as he saw a line of disturbance in the water which appeared suddenly to be coming out from near where the other ship had stopped. It soon became a long wave which gained in speed and height. It hit the moving Zodiac at an angle and pushed it back and up against a large ice flow. Ustrashkin and his crew were violently thrown out of the boat and into the icy cold water; the Zodiac flipping completely over and landing upon the flow. The men struggled in the cold water for a while, one raising his arm for help, but then he too disappeared below the still water which had already began to take on that dull greasy appearance as it started to freeze.

Now frantic with anger and fear, Volkov stormed back into the bridge and went up to the Officer of the Watch. "Is there a mortar on board?" he yelled at the young man. "When I was at sea with my father, his icebreaker carried an old army mortar which was used to fire bombs ahead to clear some of the thick ice. Do you have one?" he yelled.

"Yes…sir, Mr Volkov." He stammered. I think that we have a mortar in our Secure Locker, but we have never yet had to use it."

"Get it now!" he ordered, and the young officer went to the bridge telephone, dialled number and gave some instructions. It took a few minutes and Volkov impatiently walked back and forth across the bridge. Eventually a crewman came up the gangway and entered the bridge. He carried on his back a heavy pack which contained the mortar and two long, oval shaped stubby bombs were strapped to the outside. Volkov recognised the pack from his old days in the army with his late friend Ustrashkin. It was an old RM-38, a Soviet era 50 mm light infantry mortar.

"Good! Just want I needed." He said taking the heavy pack from the crewman. It slung it onto his back and leaving the bridge rushed down the gangway and made his way to the open bow of the ship. Here he unpacked the contents of the pack, assembling the base plate which he planted firmly on the deck. He took out the small barrel which was about three quarters of a metre long and screwed it into the base. The barrel only had two elevation angles, 45 and 75 degrees. He altered the sleeve at the base of the barrel to give the maximum range at 45 degrees and pointed it towards the other ship. He took one of the bombs and removed the safety protector at its base. He got down upon his knees and sighted over the end of the barrel towards the *Mystery*. Satisfied that he had aimed the mortar accurately and that it was set on the maximum range, he thrust one of the bombs down the barrel. There was a short 'thud' and flame with a strong vibration of the deck plates as the bomb was projected out of the barrel. Volkov stood up and watched the trajectory of the slow-flying bomb towards the *Mystery*. It landed on an ice flow well in front of the other ship and exploded with a flash and shower of ice. The sound arriving a short delay afterwards.

"Short!" he swore to himself. He stood up and slung the pack and its remaining bomb onto his back and slowly lifted up the assembled mortar. He carried it along the deck to amidships where the other Zodiacs were stored. He yelled at several of the crew who were standing nearby, perplexed at the morning's dramatic events. "Get one of those things over the side!" he yelled pointing to the rack which carried the rest of the inflatables.

Before one was taken up to be lowered over the open section of the railing, Volkov carefully placed the pack and then the assembled mortar into the Zodiac. He needed more range. He remembered from his army days that this type of Mortar only had a range of about eight hundred metres. He would have to get closer to the other ship; at least another two to three hundred metres. Volkov stepped into it and it was lowered into the small section of open water which surrounded the ship. He started the outboard and turned the small craft towards the other ship, looking for a pathway of clear water between the flows. The greasy grey and cracked scum of ice on the water's surface told him that the sea had started to freeze over. He pushed the small Zodiac through the icy scum but he did not get very far but far enough he thought. The ice flows had begun to close together as the seawater froze at minus two degrees Celsius. He rammed the Zodiac hard up against a small flow and threw the pack with its bomb onto the ice. He picked up the assembled mortar from the wooden floor of the Zodiac and pushed it onto the edge of the ice. He scrambled onto the ice and it tilted slightly with the new weight on its edge. The Zodiac moved off away from the flow and into the small patch of open water. Volkov balanced himself carefully as he stood up on the swaying flow, conscious of the fact that flows of this size were often very unstable. He found an indentation in the ice which had been covered with snow and set the base of the mortar firmly into it. He knelt down, and again sited the mortar towards the *Mystery*. He was well in range now and he could not miss. He prepared the second mortar bomb and with a smirk on his face and took one last look at his target. He thrust the bomb down the short barrel. There was the usual explosion, and the bomb lazily left its tube. With a sudden feeling of fear, he realised that the angle of the mortar had changed and was now pointing higher in the sky; the ice flow was slowly tilting. Volkov grabbed the hot barrel of the mortar in a vain attempt to stop himself from sliding down the ever-tilting ice flow. It was not enough. The force of the bomb leaving the barrel which had been placed at one end of the flow had caused it to tilt suddenly. Volkov spread his body out on the surface of the flow in a vain attempt to stop his motion. But the flow had been shocked enough by the explosion to now complete its tilt until it suddenly turned over, its rugged, transparent base now pointing to the grey sky and Volkov and his mortar had disappeared below the ice and below the cold water of the Barents Sea.

Epilogue

The threat to the passengers and crew aboard the *Mystery of the Seas* and their secret mission into the Arctic Ocean was over. The ship continued on its way to Longyearbyen in Svalbard where the passengers took their various fights home. They had been totally unaware of the dangerous events which had confronted Breck and the crew, although Fraser, the marine biologist, had often thought that something had been amiss.

Captain Khovrin had his diver free the propellers of the *Vasili Brusilov* of the mysterious rope which had fouled them and returned to Murmansk where the authorities were told of the tragic death of Dimitri Volkov and his security team in trying to help another ship which appeared to have been trapped in the ice pack.

Terence Breck had been both saddened and relieved to leave the *Mystery of the Seas*. He had made good friends with the crew, especially the

First Officer, Calypso and the captain called 'Poseidon' by his crew. He had flown back to New Bedford where he had been welcomed by his wife, June and their son Tony. They had finished helping her parents recover from the freak Florida tsunami, and after a short stay had also returned home.

On returning to his duties with the U.S. Naval Oceanographic Office (NAVOCEANO), Breck found to his great delight that Poseidon had been true to his word, and that his mysterious 'organisation' and deposited several million dollars into a research fund on the proviso that it be used by Breck and NAVOCEANO to chart the bottom currents along the Norwegian coast well into the Arctic Ocean. It took several long months for this research to be organised and once Breck had begun to carry out the official version of his past research, he found that it took much longer, as it was more thorough than his original clandestine study on board the *Mystery of the Seas*. Moreover, it was also highly publicized and attracted the attention of many of the world's oceanographic and climate researchers. By the time that Breck was able to publish his research, which included the contribution of data given to him by the late Dr. Yuri Ziminov, the yearly snowfall in northern Siberia had been measured by new researchers and was found to be at record levels. Moreover, it was also noted that the winter ice pack had begun to spread more thickly across the Arctic Ocean. Breck's research and reports by NAVOCEANO to the Secretariat of the UNFCCC (the United Nations Framework Convention on Climate Change), finally prompted an international response to this threat of a new ice age. This led to the setting up of a practical sub-committee of the United Nations designed to monitor and control international heat flow into the oceans and atmosphere.

With the Arctic showing signs of returning to a normal climate and with further research showing that the Norwegian Current was also strengthening, Dr. Terence Breck and his family returned to the seaside cottage of June's parents for a well-earned holiday. Sitting out on the porch overlooking a calm, blue Atlantic, Breck was not saddened nor surprized when he read in the Shipping News that the ecotourism ship, the *Mystery of the Seas* had disappeared in the Mediterranean. Apparently, she was travelling for a refit in a shipyard in Greece and carried no passengers but only crew. She was lost somewhere near the Maltese island of Gozo, thought to be the mythical island of Ogygira, the home of the sea nymph Calypso.

Author's Note

This is a work of fiction but the science is real. I studied the 1956 theory of Maurice Ewing and William L. Donn to explain the Ice Ages when I was an Undergraduate way back in the 1970's. At that time, before global warming became well known, they could not explain why the Arctic Ocean would be free of ice thus precipitating an Ice Age. I have included some of the most recent published papers from ethical scientific journals but I have also travelled to all seven continents making my own observations.

The crew of the ship *Mystery of the Seas* are based on ancient Greek and Celtic myths:

Poseidon – who presided over the sea, storms, earthquakes, horses and was the protector of seafarers. The captain's full name, Filip Athanasios, means "immortal - friend of horses";

Calypso - was a nymph who lived on the island of Ogygira (now Gozo in Malta), a daughter of Oceanus and Tethys. The First Officer's surname, Thalassa means "personification of the sea";

Damon – was a disciple of the great physician Asclepius. His name symbolizes the ability to tame and control illnesses;

Armenus - the Russian-speaking Chief Steward was an Argonaut who travelled with Jason into the Black Sea to find the Golden Fleece; and

Fionn and **Rónán** – who helped Breck with the oceanographic gear and are the cold water divers of the ship. The names are of Celtic origin, meaning "fair" and "seal" and both are Selkies (Seal People) from the northern isles of Scotland.

About the Author

Dr Peter Scott the author, like Breck the main character of this novel, has also been trained in oceanography and has travelled down the coasts of the Antarctic Peninsular and Norway as well as to many other parts of the world. He has had an award-winning teaching career, mainly in the Earth Sciences, been an Officer in the Australian Army Reserve, an Officer-Instructor in the Australian Naval Cadets and has trained under sail in Tall Ships. He has written over thirty books in science and other genres, including twelve novels.

www.ingramcontent.com/pod-product-compliance
Lightning Source LLC
Chambersburg PA
CBHW070953180726
48291CB00004B/1274